Springer Theses

Recognizing Outstanding Ph.D. Research

Aims and Scope

The series "Springer Theses" brings together a selection of the very best Ph.D. theses from around the world and across the physical sciences. Nominated and endorsed by two recognized specialists, each published volume has been selected for its scientific excellence and the high impact of its contents for the pertinent field of research. For greater accessibility to non-specialists, the published versions include an extended introduction, as well as a foreword by the student's supervisor explaining the special relevance of the work for the field. As a whole, the series will provide a valuable resource both for newcomers to the research fields described, and for other scientists seeking detailed background information on special questions. Finally, it provides an accredited documentation of the valuable contributions made by today's younger generation of scientists.

Theses may be nominated for publication in this series by heads of department at internationally leading universities or institutes and should fulfill all of the following criteria

- They must be written in good English.
- The topic should fall within the confines of Chemistry, Physics, Earth Sciences, Engineering and related interdisciplinary fields such as Materials, Nanoscience, Chemical Engineering, Complex Systems and Biophysics.
- The work reported in the thesis must represent a significant scientific advance.
- If the thesis includes previously published material, permission to reproduce this must be gained from the respective copyright holder (a maximum 30% of the thesis should be a verbatim reproduction from the author's previous publications).
- They must have been examined and passed during the 12 months prior to nomination.
- Each thesis should include a foreword by the supervisor outlining the significance of its content.
- The theses should have a clearly defined structure including an introduction accessible to new PhD students and scientists not expert in the relevant field.

Indexed by zbMATH.

Keita Mori

Metal-Responsive Base Pair Switching of Ligand-type Uracil Nucleobases

Doctoral Thesis accepted by
The University of Tokyo, Tokyo, Japan

Springer

Author
Dr. Keita Mori
Department of Applied Chemistry,
Graduate School of Engineering
Tokyo University of Agriculture
and Technology
Koganei, Tokyo, Japan

Supervisor
Prof. Mitsuhiko Shionoya
Department of Chemistry, Graduate School
of Science
The University of Tokyo
Bunkyo, Tokyo, Japan

ISSN 2190-5053 ISSN 2190-5061 (electronic)
Springer Theses
ISBN 978-981-99-9399-4 ISBN 978-981-99-9400-7 (eBook)
https://doi.org/10.1007/978-981-99-9400-7

This Springer imprint is published by the registered company Springer Nature Singapore Pte Ltd.
The registered company address is: 152 Beach Road, #21-01/04 Gateway East, Singapore 189721, Singapore

Supervisor's Foreword

DNA is a molecule responsible for storing and transmitting genetic information, and its mechanistic basis in expressing its functions is the association and dissociation of nucleic acid base pairs. In nature, hydrogen bonds formed between adenine (A) and thymine (T), and between guanine (G) and cytosine (C), are the key to base pairing. Hydrogen-bonded base pairing determines the thermal stability of double-stranded and triple-stranded DNA in a sequence-specific manner, making it possible to construct dynamic systems such as strand displacement.

Our group began research on "artificial metallo-DNA" in the late 1990s, focusing on "metal-mediated base pairing" in DNA duplexes. Since the first example in 1999, we have reported the design and synthesis of various metal-mediated base pairs and have demonstrated metal array in DNA duplexes, functional switching of DNAzymes, and strand displacement-based molecular machines.

As part of this, Dr. Keita Mori produced two important research results during his doctoral course: (1) DNA hybridization partner switching system using a synthetic nucleobase that has both hydrogen bonding ability and metal complex formation ability and (2) metal-mediated DNA strand displacement and the first application of metal-mediated base pair switching to molecular device manipulation.

This study greatly expands the possibilities of metal-mediated base pairing, and the achievement was published in *Chemical Science* (2023) and *Nature Communications* (2023). Dr. Keita Mori was also awarded Springer Nature Poster Award in the 15th International Symposium on Applied Bioinorganic Chemistry, CSJ Student Presentation Award in the 101st Annual Meeting of the Chemical Society of Japan, and Research Encouragement Award from Graduate School of Science, The University of Tokyo. I strongly hope that Dr. Mori's great effort and research results on artificial DNA based on metal-mediated base pairing will contribute to the further development of nucleic acid synthesis chemistry and DNA nanotechnology.

Tokyo, Japan
November 2023

Prof. Mitsuhiko Shionoya

Acknowledgements

This research was promoted under supervision of Professor Dr. Mitsuhiko Shionoya (The University of Tokyo). I would like to express my deepest appreciation to him for his kind guidance and thoughtful advice. His passion for the pursuit of science has kept encouraging me to work on challenging projects throughout my life in his laboratory. Every discussion with him, including casual talks especially about basketball, has made me full of energy to tackle any difficulties in my research. It has been a great honor for me to be a member of the Shionoya Lab family.

I sincerely appreciate tremendous support from Assistant Professor Dr. Yusuke Takezawa (The University of Tokyo) for my research. He has always been a perfect role model as a scientist for me. Countless discussions I had with him will continue to be my invaluable memory. I will never forget his smile and excitement when I let him know any tiny progress of my research. Without his technical assistance and professional feedbacks, I could never have carried out any projects in my Ph.D. course.

I am grateful to Associate Professor Dr. Shohei Tashiro and Assistant Professor Dr. Hitoshi Ube (The University of Tokyo) for their continuous support. Their suggestions from different viewpoints have been helpful for me to accomplish my Ph.D. course study.

I would like to acknowledge insightful advice on my research from Assistant Professor Dr. Zhen Lei (Fuzhou University) and Project Assistant Professor Dr. Xiao-Li Pei (The University of Tokyo).

I thank Dr. Koichi Nagata (Tohoku University) and Dr. Jens Christian Léon for their kind encouragement and guidance. I wish we could have worked together for longer period.

I am deeply grateful to Dr. Manabu Nakaya (Josai University) and Dr. Takahiro Nakama (The University of Tokyo) for raising me as a scientist. I believe that we definitely shared a special time in Takezawa group. I have learnt a lot of important things as a professional researcher during the time with them.

I am thankful to Dr. Kotaro Nishiyama (Meiji University), who started a research project regarding the 5-hydroxyuracil base and left invaluable knowledge and experimental data in Shionoya laboratory. It has been a great honor for me to take over his project, and I really appreciate his kind backup.

I thank all the members who shared our life in Shionoya laboratory. Without their support, I could never have survived my Ph.D. program. I would like to express my gratitude especially to Ms. Shiori Sakakibara, Mr. Akira Suzuki, Ms. Tong Xing, Dr. Shuntaro Amano, Mr. Nobuyuki Fuse, Dr. Lingyun Hu, Dr. Chandran Rajasree Silpa, Mr. Daisuke Kanemaru, Mr. Naofumi Kudo, Ms. Wei-En Huang, Mr. Masahiro Wakano, Ms. Hanci Zhang, Mr. Kohei Sakurai, Mr. Sentaro Obana, and Mr. Hiroshi Kasuga, who shared our research in Takezawa group. I have enjoyed every discussions with them, through which my understanding of our research has been greatly deepened. I am also grateful to Dr. Yutaro Yamashita, Dr. Ryunosuke Hayashi, Dr. Ayako Hino, and Mr. Tomoki Nakajima for sharing our time as Ph.D. course students. Their diligent attitudes and painstaking researches have motivated me throughout my journey in Shionoya laboratory.

I am grateful to Professor Dr. Hanadi Sleiman (McGill University) and the colleagues in her laboratory for kindly accepting me as an internship student. The stay in her laboratory and the beautiful city of Montreal has been a special part of my Ph.D. course.

I am thankful to Dr. Shinjiro Takano and Mr. Yoshimitsu Hori (The University of Tokyo) for their technical support regarding ESI- and MALDI-TOF MS measurements.

I would like to express my gratitude to the Japan Society for the Promotion of Science (JSPS) and Global Science Graduate Course (GSGC) for their financial support. I appreciate fruitful discussions with Professor Dr. Takeaki Ozawa (The University of Tokyo) as my secondary supervisor of the GSGC program.

Finally, I would like to express my heartfelt gratitude to my family, Dr. Yukishige Ito, Dr. Natsuko Ito, and Dr. Kohta Ito for their warm support, constant encouragement, and love in my life.

Keita Mori

Contents

List of Published Articles

Mori, K.; Takezawa, Y.; Shionoya, M. Metal-dependent base pairing of bifacial iminodiacetic acid-modified uracil bases for switching DNA hybridization partner. *Chem. Sci. 2023*, **14**, 1082–1088.

Takezawa, Y.; Mori, K.; Huang, W.-E.; Nishiyama, K.; Xing, T.; Nakama, T.; Shionoya, M. Metal-mediated DNA strand displacement and molecular device operations based on base-pair switching of 5-hydroxyuracil nucleobases. *Nat. Commun. 2023*, **14**, 4759.

Abbreviations

8HQ	8-hydroxyquinoline
A	Adenine
A, Abs.	Absorbance
a.u.	Arbitrary unit
Ac	Acetyl
AcOH	Acetic acid
AFM	Atomic force microscopy
AP	2-aminopurine
ATP	Adenosine triphosphate
BAPTA	*O*,*O*′-bis(2-aminophenyl)ethyleneglycol-*N*,*N*,*N*′,*N*′-tetraacetic acid
bp	Base pair(s)
bpy	Bipyridine
C	Cytosine
calcd.	Calculated
caU	5-carboxyuracil
CD	Circular dichroism
Dabcyl	4-(4-dimethylaminophenylazo)benzoic acid
dcaU	*N*,*N*-dicarboxymethyl-5-aminouracil
DIPEA	Diisopropylethylamine
Dipic	Pyridine-2,6-dicaboxylate
DMAP	4-dimethylaminopyridine
DMTr	4,4′-dimethoxytrityl
DNA	Deoxyribonucleic acid
dpp	Diphenyl phenanthroline
dsDNA	Double-stranded DNA
DX	Double-crossover
EDTA	Ethylenediaminetetraacetic acid
EGTA	Ethylene glycol bis(2-aminoethyl ether)-*N*,*N*,*N*′,*N*′-tetraacetic acid
equiv.	Equivalent(s)
ESI	Electrospray ionization
FAM	6-carboxyfluorescein

FRET	Fluorescence resonance energy transfer
G	Guanine
H	Hydroxypyridone
HCR	Hybridization chain reaction
HEPES	4-(2-hydroxyethyl)-1-piperazineethanesufonic acid
HPA	Hydroxypicolinic acid
HPLC	High-performance liquid chromatography
HRMS	High-resolution mass spectrometry
I	Fluorescence intensity
IDA	Iminodiacetic acid
Im	Imidazole
$\mathbf{Im}^{\mathrm{C}}$	4-carboxyimidazole
^{i}Pr	Isopropyl
K_{ML}	Stability constant of a metal–ligand complex
M	Molar/metal ion
m/z	Mass-to-charge ratio
MALDI	Matrix-assisted laser desorption ionization
MeCN	Acetonitrile
MeOH	Methanol
MS	Mass spectrometry
N	Natural nucleobases
NMR	Nuclear magnetic resonance
nt	Nucleotide(s)
NV	6-nitroveratryl
obs.	Observed
PAGE	Polyacrylamide gel electrophoresis
Phos	5′-phosphate
PNK	Polynucleotide kinase
py, **Py**	Pyridine
RNA	Ribonucleic acid
S	Salen
SB	4-hydroxy-2-mercaptobenzimidazole
SDR	Strand displacement reaction
ssDNA	Single-stranded DNA
T	Thymine
$t_{1/2}$	Half-completion time
TEAA	Triethylammonium acetate
TLC	Thin-layer chromatography
T_{m}	Melting temperature
TMS	Tetramethylsilane
TOF	Time-of-flight
tpy	Terpyridine
Tris	Tris(hydroxymethyl)aminomethane
U	Uracil
$\mathbf{U}^{\mathbf{OH}}$	5-hydroxyuracil

UV	Ultraviolet
$\mathbf{U^X}$	5-modified uracil
vis	Visible
z	Charge
ε	Molar extinction coefficient
λ_{em}	Emission wavelength
λ_{ex}	Excitation wavelength

Chapter 1
General Introduction

Abstract Sequence-dependent self-assembly of DNA molecules has been used to develop advanced nanoarchitectures, giving rise to "DNA nanotechnology". Since Seeman et al. reported branched DNA assemblies, methodologies for fabricating DNA nanostructures have been developed dramatically, represented by DNA origami technique. Dynamic conversion of DNA structures has also been achieved based on strand displacement reactions (SDRs), resulting in DNA-based molecular machines and nanodevices operated by specific sequences of oligonucleotides. Recently, it has gained much attention to construct DNA systems driven by external stimuli instead of nucleic acid strands. In particular, metal ions show unique properties such as reversible complexation, stimuli-responsiveness, and binding specificity. This chapter overviews the history of DNA nanotechnology spanning from early studies of DNA nanostructures to recent examples of dynamic DNA nanodevices. Stimuli-responsive DNA systems are also highlighted, focusing on characteristics of each stimulus. In particular, metal-mediated base pairing is discussed as a strategy for metal-responsive control of DNA assemblies.

Keywords DNA nanotechnology · DNA strand displacement reaction · Stimuli-responsive control · Metal-mediated base pairing · Modified uracil

1.1 Programmable Self-Assembly of DNA

A DNA molecule is one of the essential biopolymers in living systems, normally conceived as a carrier of genetic information. The DNA strand is a polymer of deoxyribonucleotides, each of which consists of a phosphodiester, a deoxyribose, and a nucleobase (Fig. 1.1a). Four natural nucleobases, adenine (A), thymine (T), guanine (G), and cytosine (C), recognize their counterparts and selectively form Watson–Crick-type base pairs, A–T and G–C, through hydrogen bonding (Fig. 1.1b). This hydrogen-bonded base pairing allows for the formation of an antiparallel double helix which consists of two DNA strands with complementary base sequences. When a mismatch other than the A–T or G–C pair was introduced, DNA duplexes are greatly

K. Mori, *Metal-Responsive Base Pair Switching of Ligand-type Uracil Nucleobases*, Springer Theses, https://doi.org/10.1007/978-981-99-9400-7_1

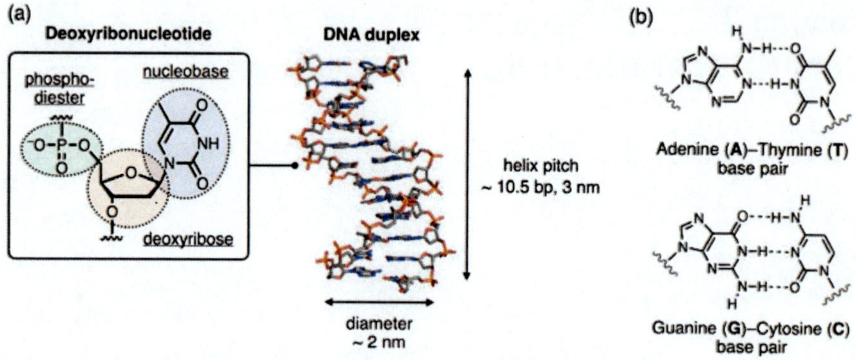

Fig. 1.1 **a** Structures of a deoxyribonucleotide and DNA duplex. **b** Watson–Crick hydrogen-bonded base pairs of natural nucleobases

destabilized. Thus, DNA shows sequence-specific self-assembly to form a thermodynamically most stable structure under given conditions. Solid-phase DNA synthesis based on phosphoramidite chemistry has been established as the most common method to obtain oligonucleotides with defined length and sequences. Therefore, DNA has gathered increasing attention not only as a functional biomolecule, but also as a building block of nanomaterials.

Since DNA has well-defined double helices and shows sequence-dependent hybridization, self-assembly of DNA strands is highly predictable and controllable. Researchers have developed computational tools to predict secondary structures of DNA strands based on the base sequences [1, 2]. Accordingly, various DNA-based nanoarchitectures have been developed so far, giving rise to a research field called "DNA nanotechnology" [3–7]. As a pioneering work, Seeman et al. constructed two-dimensional branched networks and three-dimensional cubic nanostructures by self-assembly of several short DNA strands (Fig. 1.2a) [8–11]. Turberfield et al. reported self-assembly of four DNA strands to build "DNA tetrahedra" (Fig. 1.2b) [12, 13]. Seeman et al. subsequently reported double-crossover (DX) tiles formed by hybridization of five DNA strands (Fig. 1.3a) [14, 15]. Because the DX tiles are relatively robust compared with simple DNA duplexes, they have been utilized as a building block of various higher-ordered nanoarchitectures [16]. The DX tiles can be assembled through the base-pairing interactions in short single-stranded regions at their termini, termed as "sticky ends". For example, Yan and LaBean et al. constructed two-dimensional networks using four-way branched DNA tile monomers (Fig. 1.3b) [17]. Mao et al. utilized three-way-junction tile monomers for building two-dimensional hexagonal networks [18, 19] (Fig. 1.3c) and three-dimensional polyhedral structures [20] (Fig. 1.3d). Furthermore, it was achieved to fabricate three-dimensional DNA crystals with a micrometer-scale size (Fig. 1.3e) [21].

In 2006, Rothemund reported "DNA origami" as a new technique for constructing DNA nanostructures of precisely defined size and shape [22]. In this method, a long

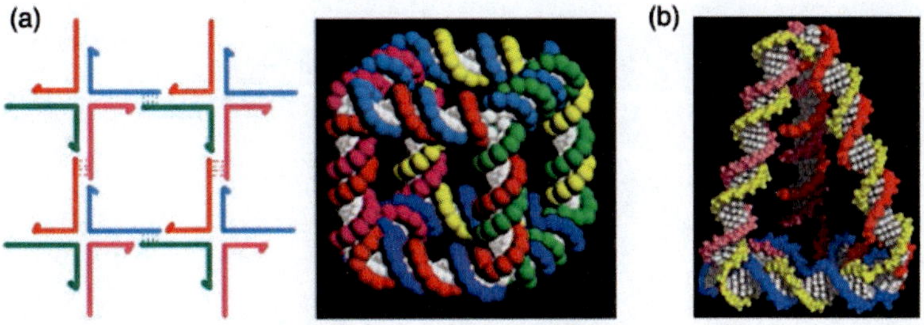

Fig. 1.2 Two- and three-dimensional assembly of DNA duplexes. **a** Branched DNA assemblies and DNA cubes. **b** DNA tetrahedra. Adapted with permission from Ref. [6]. Copyright 2019 American Chemical Society

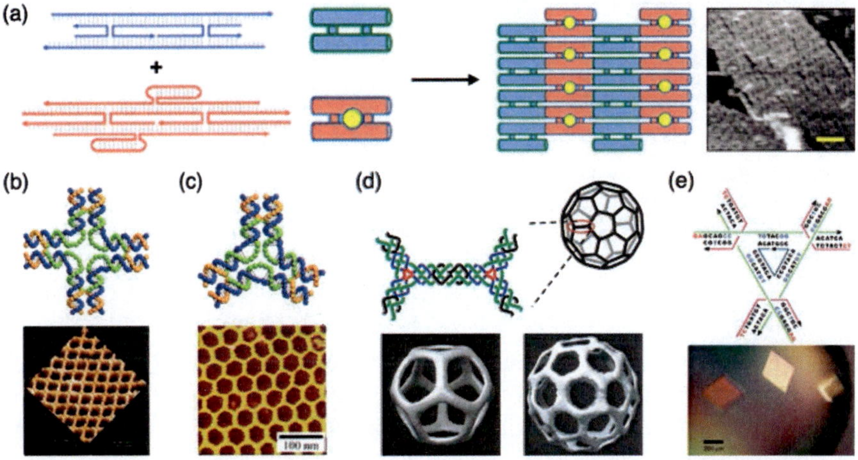

Fig. 1.3 DNA nanostructures based on DNA double-crossover (DX) tiles. **a** Two-dimensional assembly of DX tiles. Adapted with permission from Ref. [7]. Copyright 2019 American Chemical Society. **b** Two-dimensional networks formed by four-way branched DNA tiles. Adapted with permission from Ref. [23]. Copyright 2017 American Chemical Society. **c** Two-dimensional hexagonal assembly of three-way branched DNA tiles. Adapted with permission from Ref. [23]. Copyright 2017 American Chemical Society. **d** DNA polyhedra constructed by three-way branched DNA tiles. Adapted with permission from Ref. [6]. Copyright 2019 American Chemical Society. **e** Three-dimensional DNA crystals. Adapted with permission from Ref. [6]. Copyright 2019 American Chemical Society

"scaffold strand" is folded into a specific shape by the hybridization of many "staple strands" (Fig. 1.4a). Various nanostructures including even asymmetric shapes were successfully created using the DNA origami technique [23, 24]. Furthermore, three-dimensional DNA boxes [25] (Fig. 1.4b) and more complicated asymmetric architectures [26] (Fig. 1.4c) were also fabricated. In addition, helix bundles formed by the parallel assembly of interconnected DNA duplexes have emerged as a robust structural motif (Fig. 1.4d) [27, 28]. They have recently been utilized as the body

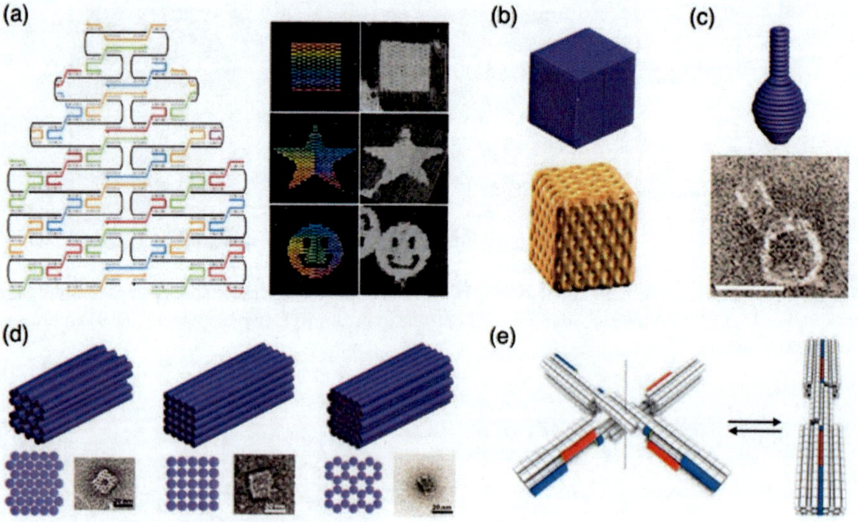

Fig. 1.4 DNA origami-based nanostructures. **a** Design and two-dimensional assembly of DNA origami structures. Adapted with permission from Ref. [6]. Copyright 2019 American Chemical Society. **b** Cube and **c** nanoflask constructed by the DNA origami technique. Adapted with permission from Ref. [23]. Copyright 2017 American Chemical Society. **d** DNA helix bundles. Adapted with permission from Ref. [23]. Copyright 2017 American Chemical Society. **e** DNA nanodevices based on the helix bundles. Adapted from Ref. [30]. Copyright 2020 American Chemical Society

of DNA-based nanodevices and mechanical motors (Fig. 1.4e) [29–33]. It is also important to note that modeling software such as caDNAno [34] makes it easy to design DNA origami assemblies with specific morphologies.

Chemical modification of the DNA nanostructures greatly expands their functions [35]. The modification methods have been widely studied to present several chemical reactions as efficient labeling approaches (Fig. 1.5a). Gothelf et al. demonstrated that various functional molecules can be immobilized on a DNA origami surface by performing orthogonal chemical reactions in a stepwise manner (Fig. 1.5b) [36]. Since proteins can be introduced into the termini of DNA strands, the immobilization of two different enzymes on the surface of DNA nanostructures was also achieved (Fig. 1.5c) [37]. It was revealed that the target reaction was significantly enhanced by the enzyme cascade when the distance between the two enzymes was optimal. In addition, assembly of functional molecules along DNA duplexes has been achieved. Liu et al. reported that three different dye molecules can be incorporated in DNA helix bundles (Fig. 1.5d) [38]. The light-harvesting ability was enhanced by accumulating the dyes along the DNA scaffolds. Hybrids of DNA assemblies and biomolecules have been constructed by introducing target-binding functionalities into DNA nanostructures. For example, cholesterol-modified helix bundles were inserted into lipid bilayers as synthetic lipid membrane channels (Fig. 1.5e) [39].

Thus, DNA nanostructures have been regarded as an excellent building block of functional materials for the following reasons. (i) Their shape and size are highly

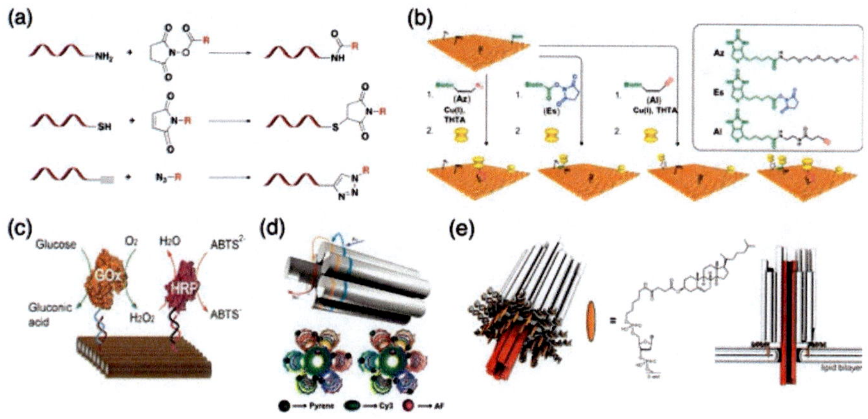

Fig. 1.5 DNA nanostructures modified with chemical functionalities. **a** Reactions to introduce functional molecules into the termini of DNA strands. Adapted with permission from Ref. [35]. Copyright 2019 American Chemical Society. **b** Immobilization of different functionalities on DNA origami surfaces by orthogonal chemical reactions. Adapted with permission from Ref. [35]. Copyright 2019 American Chemical Society. **c** Cascade enzymes immobilized on the surface of DNA nanostructures. Adapted with permission from Ref. [37]. Copyright 2012 American Chemical Society. **d** Light-harvesting antenna constructed by immobilizing multiple dye molecules into DNA helix bundles. Adapted with permission from Ref. [38]. Copyright 2011 American Chemical Society. **e** Synthetic lipid membrane constructed by cholesterol-modified DNA nanostructures. Adapted with permission from Ref. [6]. Copyright 2019 American Chemical Society

programmable and predictable based on well-established computational algorithms such as NUPACK and caDNAno [1, 2, 34]. (ii) Synthetic methods for DNA strands and nanostructures are well established. (iii) DNA structures are compatible with additional modification by chemical handles. DNA nanotechnology is still rapidly growing as an attractive research field spanning from supramolecular chemistry to biotechnology and materials engineering.

1.2 Dynamic Control of DNA Assemblies Based on Strand Displacement Reactions

DNA molecules have been utilized to develop stable nanostructures based on their sequence-dependent self-assembly. Furthermore, dynamic conversion of their structure has also been achieved, featured as "dynamic DNA nanotechnology" [40, 41]. Dynamic DNA nanotechnology largely depends on toehold-mediated DNA strand displacement reactions (SDRs) (Fig. 1.6a) [42]. In this process, one strand of an original DNA duplex is replaced by an input strand to form another duplex with higher thermal stability. The input strand binds to a terminal single-stranded region of the original duplex, termed as "toehold". After the toehold binding at one end, the branching point of the three-stranded complex gradually migrates to the other

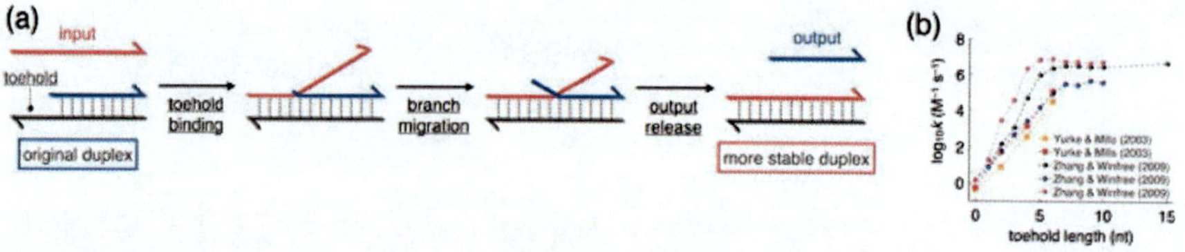

Fig. 1.6 Toehold-mediated DNA strand displacement reactions (SDRs) using natural DNA strands. **a** Mechanism of the toehold-mediated SDRs. **b** Effects of the toehold length on the speed of the SDRs. Adapted with permission from Ref. [42]. Copyright 2019 American Chemical Society

terminus via "branch migration". Finally, the branch migration completes to release an output strand, yielding a desired full-match duplex. It has been reported that the rate of the toehold-mediated SDRs is exponentially enhanced as the length of the toehold region is increased, and almost maximized when the length reaches 6 nt (Fig. 1.6b) [43, 44].

The toehold-mediated SDRs have been widely applied to create dynamic DNA molecular machines and nanodevices [45, 46]. As a pioneering work, Yurke et al. developed DNA tweezers (Fig. 1.7a) [47]. The "closed" state of the tweezers can be formed by self-assembly of four DNA strands. The addition of an input strand induces the SDR to remove the closing strand, yielding the open form. The closed tweezers can be regenerated by adding the closing strand again. Pierce et al. applied the SDR to the regulation of DNA molecular machine showing a progressive motion, termed as a "DNA walker" (Fig. 1.7b) [48, 49]. The DNA duplex acts as a walker that moves along the track by the sequential addition of different input strands, which induce cascade SDRs. Mao et al. developed a "DNA gear", in which the relative position of the two circular DNAs can be dynamically controlled using SDRs (Fig. 1.7c) [50]. Furthermore, SDRs are applicable to dynamic nanodevices composed of higher-ordered nanostructures. For instance, a DNA origami-based container can be opened by the addition of an input strand, resulting in the activation of an encapsulated enzyme (Fig. 1.7d) [51]. In addition, a Möbius strip, a topological ribbon-like structure, was constructed by the DNA origami technique (Fig. 1.7e) [52]. It was demonstrated that the Möbius strip can be converted to a loosened ring by SDRs removing base-pairing interactions between staple strands and the central scaffold strand.

The SDRs are also applicable to various DNA structural motifs other than simple duplexes. One prominent example is hybridization chain reactions (HCRs) using DNA hairpins (Fig. 1.8a) [53–55]. When an initiator strand is added to a mixture of two DNA hairpins 1 and 2, hairpin 1 can be opened by the toehold-mediated SDR. As a result, a toehold for the subsequent SDR with hairpin 2 is exposed. Thus, the cascade SDRs continuously open the two hairpins, yielding long DNA duplexes. The HCRs have been broadly utilized especially for amplification of input signals [56]. For example, Huang et al. utilized a pyrene-modified hairpin and cationic β-CD polymer for detection of a target DNA strand [57]. Without the target strand, fluorescence of the pyrene is enhanced due to the encapsulation by β-CD. When

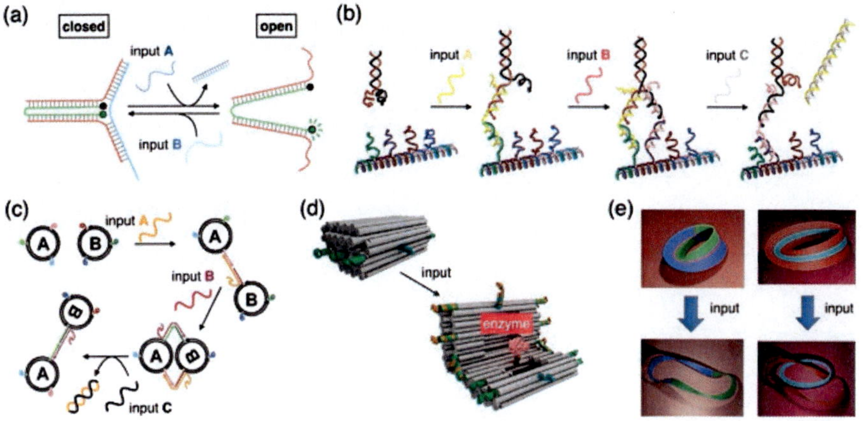

Fig. 1.7 DNA-based molecular machines and nanodevices controlled by toehold-mediated strand displacement reactions (SDRs). **a** DNA tweezers. Adapted with permission from Ref. [71]. Copyright 2014 American Chemical Society. **b** DNA walker. Adapted with permission from Ref. [48]. Copyright 2004 American Chemical Society. **c** DNA gear. Adapted with permission from Ref. [50]. Copyright 2004 American Chemical Society. **d** Opening of a DNA container with internal functionalization by an enzyme. Adapted with permission from Ref. [42]. Copyright 2019 American Chemical Society. **e** Topological control of DNA Möbius strip. Adapted with permission from Ref. [33]. Copyright 2014 American Chemical Society

the HCR proceeded in the presence of the target strand and a long DNA duplex is formed, the β-CD polymer can no longer bind to the pyrene molecules, resulting in diminished emission (Fig. 1.8b). It was also demonstrated that HCRs in the cross-linking duplexes of DNA hydrogels can induce significant swelling of the hydrogel (Fig. 1.8c) [58, 59].

Multiple SDRs can be successively operated using the output of one SDR as the input of another SDR. The cascade SDRs have been utilized to construct various reaction circuits. Zhang and Winfree et al. reported an SDR network cycle using a short input strand as a "catalyst" to efficiently amplify the input signal (Fig. 1.9a) [60]. In addition, Winfree et al. fabricated an SDR-based system in which the output strand is released only in the coexistence of two input strands, exhibiting the AND-gate response (Fig. 1.9b) [61]. Inspired by this original work, various types of logic circuits have been reported, giving rise to "DNA computation" [62–65]. It has recently been reported that temporally programmed pulse signals can be generated by combination of SDRs and enzymatic reactions (Fig. 1.9c) [66].

As described above, dynamic DNA nanotechnology greatly depends on the SDRs. Due to their sequence specificity and high modularity, various sophisticated systems exhibiting dynamic functions have been demonstrated using the SDRs. However, available inputs for the SDRs and the resultant dynamic nanodevices have been limited to DNA or RNA strands in the standard toehold-mediated mechanisms. With a view to expanding the applicability of dynamic DNA nanotechnology, it is of great significance to use external stimuli instead of oligonucleotide binding as a trigger

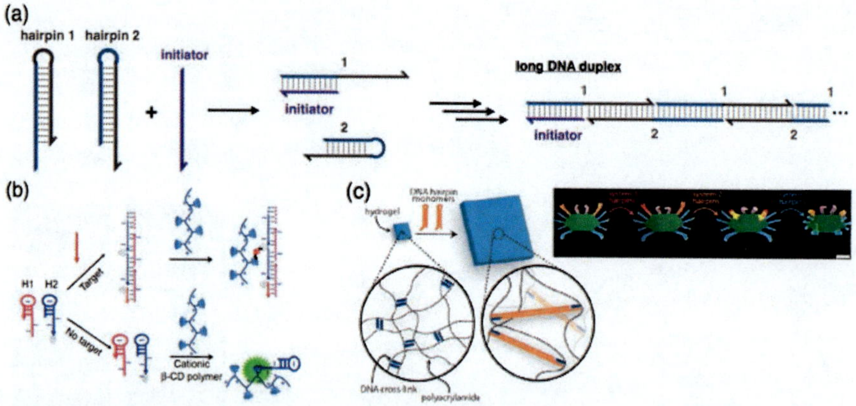

Fig. 1.8 Working principle and applications of hybridization chain reactions (HCRs). **a** Scheme of the HCRs. **b** Detection of target DNA strands with specific sequences based on enhancement of pyrene fluorescence by complexation with poly-β-CD. Adapted with permission from Ref. [55]. Copyright 2018 American Chemical Society. **c** Swelling of DNA hydrogels by HCRs. Adapted with permission from Ref. [42]. Copyright 2019 American Chemical Society

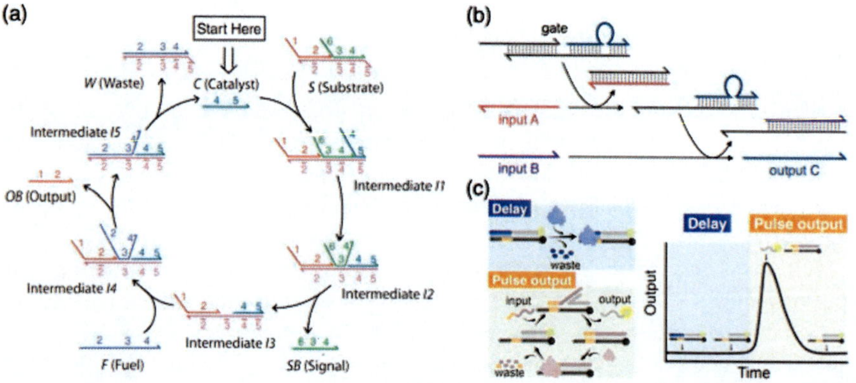

Fig. 1.9 Strand displacement reaction (SDR) circuits. **a** An SDR cycle for amplification of the input signals driven by catalyst strands. Adapted with permission from Ref. [42]. Copyright 2019 American Chemical Society. **b** DNA-based AND gate using cascade SDRs. **c** Timed pulses yielded by combination of SDRs and enzymatic reactions. Adapted with permission from Ref. [66]. Copyright 2023 American Chemical Society

of DNA structural change. Therefore, other stimuli-responsive tools to induce the SDRs and dynamic structural conversion of DNA assemblies have been demanded.

1.3 Dynamic DNA Systems Driven by External Stimuli

With a view to expanding the scope of dynamic DNA nanotechnology, it has been attempted to control DNA assemblies in response to external stimuli other than oligonucleotide binding [67, 68]. This section overviews several types of major external stimuli used for dynamic DNA systems, which are summarized in Table 1.1. Light irradiation is one of the most studied stimuli for controlling DNA structures [69–71]. By introducing artificial light-responsive unit into DNA strands, their hybridization can be reversibly controlled. pH changes have also attracted attention as external stimuli for dynamic DNA systems [72]. Dynamic conversion of DNA assemblies can be induced by pH-responsive formation of secondary structures such as DNA triplexes and i-motifs. Small molecules can also be a trigger by utilizing aptamer sequences [73]. Furthermore, metal ions have gathered substantial attention as the input of dynamic supramolecules including DNA [74–79]. Reversible and ligand-specific nature of metal complexation has broadened the programmability of dynamic DNA-based systems.

Light irradiation has been an excellent trigger for dynamic control of DNA structures because it is spatiotemporally controllable and no additional chemical species are required [69–71]. There have been several reports on photoresponsive control of DNA hybridization and strand displacement reactions (SDRs). Obika et al. introduced 4-hydroxy-2-mercaptobenzimidazole with a photocleavable 6-nitroveratryl group (**SB**$^{\mathbf{NV}}$) into DNA strands for light-responsive SDRs (Fig. 1.10a) [80]. **SB**$^{\mathbf{NV}}$ forms a hydrogen-bonded base pair with a natural guanine (G). After the photocleavage of the **NV** group, the hydrogen-bonding pattern of **SB** is changed to exhibit a base pair with an adenine (A). Accordingly, the hybridization partner of the **SB**$^{\mathbf{NV}}$-containing strand was switched by photoirradiation to induce the SDRs. Azobenzenes have also been extensively utilized to endow DNA strands with photoresponsiveness [81]. The *trans* form of azobenzene on a DNA strand can be stacked with nucleobases

Table 1.1 Features of major external stimuli utilized for dynamic DNA-based systems

Type of stimuli	Major driving force	Reversibility	Additional effect	Sequence limitation
Light irradiation	Photoisomerization of azobenzenes	Generally reversible	Photodamage of nucleobases	Introduction of many synthetic functionalities
pH change	Formation of triplexes or i-motif	Generally reversible	Effects on chemical properties of other species	Triplexes or i-motif sequences
Small molecule	Aptamer–target interaction	Generally irreversible	Not significant	Aptamer sequences
Metal ion	Ligand–metal coordination	Generally reversible	Not significant	Introduction of metal-binding ligands

of a complementary strand, resulting in the formation of a stable duplex. In contrast, after the photoisomerization, the *cis* form can no longer show the stacking interaction and the duplex is greatly destabilized. The photoresponsive hybridization of the azobenzene-modified DNA strands has been applied to reversible SDRs (Fig. 1.10b) [82–84]. Furthermore, dynamic DNA nanodevices that operate in response to photoirradiation have been reported, as represented by origami-based nanoscissors [85] (Fig. 1.10c) and switchable DNA rotaxanes [86, 87] (Fig. 1.10d). As described above, the light irradiation has been broadly utilized in dynamic DNA nanotechnology. Since the photoisomerization of azobenzenes is reversible, it is highly suitable for iterative switching of DNA structures. However, it has been known that UV light can cause damages on natural nucleobases such as photo-induced dimerization of T bases [88]. Although researchers have been making substantial efforts to develop molecular units that respond to visible light [89, 90], it is still challenging to completely avoid the photodamage and achieve high switching efficiency. In addition, introduction of many synthetic functionalities such as azobenzenes into DNA strands is required.

pH changes have also attracted attention as external stimuli for dynamic DNA systems [72]. In order to achieve pH-responsive control, DNA triplexes involving nucleobase triplets have been utilized (Fig. 1.11a). It has been known that the thymine (T) and protonated cytosine (C^+) bases can form base triplets with the A–T and G–C pairs, respectively, through Hoogsteen-type hydrogen bonding. Since the pK_a value

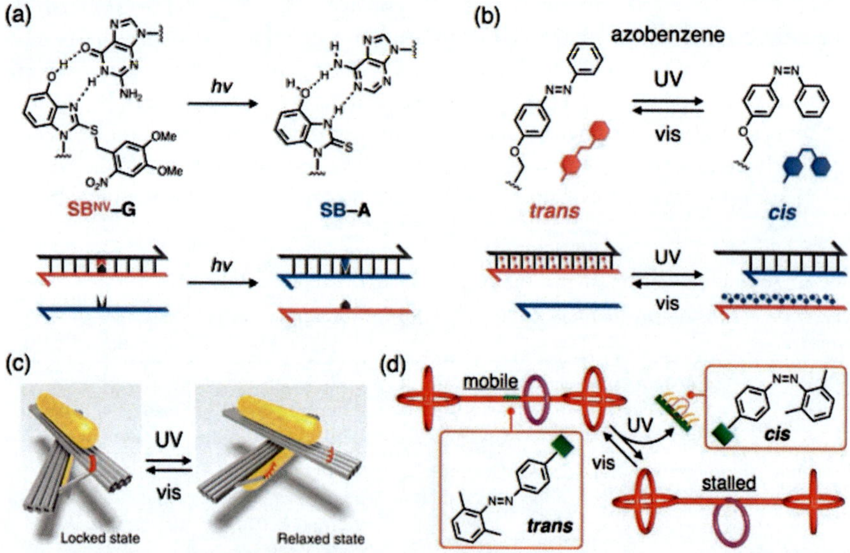

Fig. 1.10 Photoresponsive dynamic DNA systems. **a** DNA strand displacement reactions (SDRs) based on light-responsive switching of the base-pairing partner of 4-hydroxy-2-mercaptobenzimidazole (**SB**). **b** SDRs driven by photoisomerization of azobenzenes. **c**, **d** Photoresponsive dynamic DNA nanodevices driven by isomerization of azobenzenes. **c** DNA nanoscissors, **d** DNA rotaxanes. Adapted with permission from Refs. [35] and [87]. Copyright 2019 American Chemical Society and 2014 American Chemical Society

of the 3-nitrogen of cytosine base is 4.6 [91], the acidic condition is required for the formation of the C^{+}·G–C triplet. Based on the pH-dependent triplex formation, Ricci et al. developed SDRs that can be reversibly controlled by pH changes (Fig. 1.11b) [92]. Furthermore, Willner et al. applied the pH-dependent triplexes for switching the assembly pattern of DNA origami nanostructures (Fig. 1.11c) [93]. Another prominent pH-responsive nucleic acid motif is an i-motif (Fig. 1.11d) [94]. The N3-protonated cytosine (C^{+}) can form a hydrogen-bonded pair with a neutral cytosine (C) base. Therefore, a quadruplex with stacked C^{+}–C pairs inside, termed as i-motif, can be assembled around pH 5. The pH-dependent formation and dissociation of the i-motifs were utilized to control the activity of enzymes captured in a DNA-based tetrahedron (Fig. 1.11e) [95]. The RNase enzyme was activated only when the DNA tetrahedron was opened through the i-motif formation. Sol–gel transition of DNA-bridged hydrogels was also controlled by the construction of the i-motif in the crosslinking DNA strands (Fig. 1.11f) [96, 97]. Hence, pH-dependent dynamic systems have been extensively demonstrated. It is worth mentioning that pH-responsive systems can be constructed just with natural nucleobase sequences and are easily reversible. However, flexibility of the sequence design is limited since the pH-dependent assemblies such as DNA triplexes and i-motifs requires specific base sequences containing many C bases. Furthermore, since it is highly possible that pH changes affect chemical properties of other molecules involved in the solution, the pH-dependent systems would have only limited applicability.

Small molecules can also be utilized as a trigger for dynamic DNA systems using aptamer sequences [73]. Aptamer is a nucleic acid with a specific base sequence binding selectively to a target molecule, which is normally found by in vitro selection methods [98, 99]. Xu et al. reported SDRs that can be activated via exposure of the toehold region by binding of ATPs to the aptamer sequences (Fig. 1.12a) [100]. They also developed the SDRs targeting other biomolecules such as thrombin

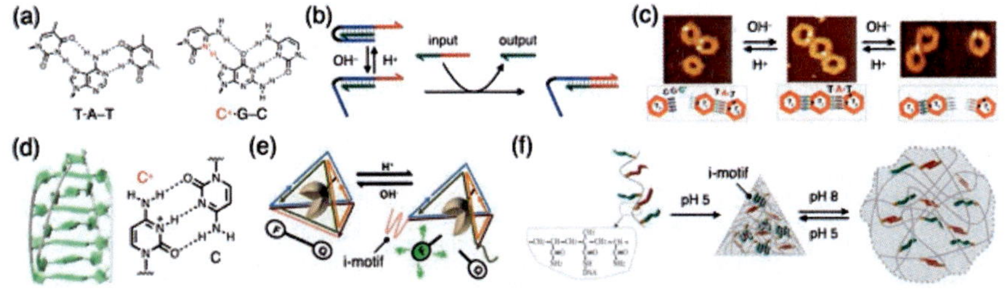

Fig. 1.11 pH-responsive dynamic DNA systems. **a** Chemical structures of nucleobase triplets. **b** pH-responsive DNA strand displacement reactions (SDRs). Adapted with permission from Ref. [92], Copyright 2014 American Chemical Society. **c** pH-dependent assembly of DNA origami nanostructures. Adapted with permission from Ref. [93]. Copyright 2016 American Chemical Society. **d** Chemical structures of the C^{+}–C pair and the resultant i-motif. Adapted from Ref. [68]. Copyright 2021 Shiliang He et al. **e** Activation of RNase A enzymes based on pH-responsive opening of DNA tetrahedra. Adapted with permission from Ref. [95]. Copyright 2017 American Chemical Society. **f** pH-dependent sol–gel transition of DNA hydrogels. Adapted with permission from Ref. [97]. Copyright 2017 American Chemical Society

proteins by introducing different types of aptamers. Dirks and Pierce demonstrated ATP-triggered HCRs based on a similar method using ATP-dependent exposure of the toeholds for the subsequent SDRs (Fig. 1.12b) [53]. This strategy was applicable to develop dynamic nanodevices using higher-ordered DNA nanostructures. For example, the morphology of DNA origami-based nanodevices called "DNA pliers" can be switched between two states (Fig. 1.12c) [101]. The duplexes bridging two domains of the DNA pliers were dissociated by ATP binding to the aptamer sequences. As a result, the "parallel" state was switched to the "cross" state in response to ATP. Although various molecules are applicable as a trigger by selecting an appropriate aptamer–target combination, it has been required to introduce long aptamer sequences into DNA strands. In addition, it is difficult to impart reversibility to the system since the aptamers show significantly high affinity to the target molecules.

Metal complexation has been widely utilized to trigger dynamic conversion of supramolecular nanoarchitectures including DNA assemblies [74–79]. It has been known that a pair of natural thymine (T) bases can form a stable 2:1 complex with a Hg^{II} ion [102, 103], so-called a "metal-mediated base pair" [104–107], inside DNA duplexes. The T–Hg^{II}–T pairing was applied to Hg^{II}-responsive SDRs (Fig. 1.13a) [108]. By introducing several T–T mismatches into the toehold sequence, the SDR proceeded only in the presence of Hg^{II} ions through toehold binding by the T–Hg^{II}–T pairing. G-quadruplexes have also been broadly used as a metal-responsive molecular tool [109]. Natural guanine (G) bases give stable planar quartets through Hoogsteen-type hydrogen bonding and coordination to the central metal ions such as K^{I}, Sr^{II}, and Pb^{II} (Fig. 1.13b). Therefore, oligonucleotides containing tandem sequences of G bases can form G-quadruplexes in the presence

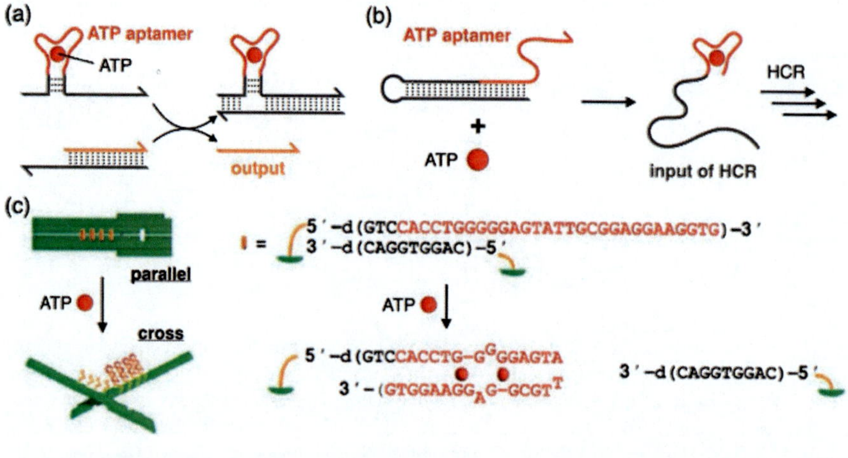

Fig. 1.12 ATP-responsive dynamic DNA systems. **a** DNA strand displacement reactions (SDRs) triggered by binding of ATPs to aptamer sequences. **b** ATP-triggered hybridization chain reactions (HCRs). **c** ATP-dependent actuation of DNA pliers. Adapted with permission from Ref. [73]. Copyright 2021 American Chemical Society

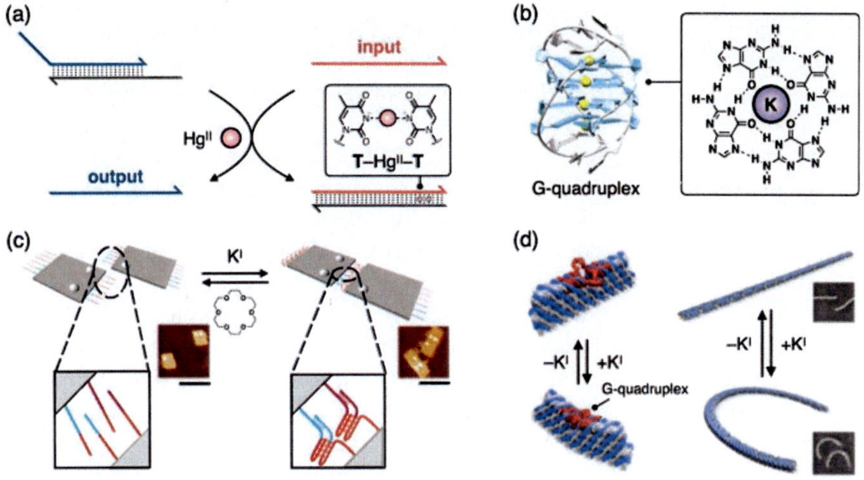

Fig. 1.13 Metal-responsive dynamic DNA systems using natural DNA strands. **a** Hg^{II}-responsive DNA strand displacement reactions (SDRs) driven by the formation of T–Hg^{II}–T base pairs. **b** Chemical structure of G-quadruplexes stabilized by K^{I} ions. Adapted from Ref. [68]. Copyright 2021 Shiliang He et al. **c** Reversible dimerization and disassembly of DNA origami constructs by the formation of G-quadruplexes. Adapted from Ref. [68]. Copyright 2021 Shiliang He et al. **d** Dynamic DNA origami nanoarms based on K^{I}-dependent formation of G-quadruplexes. Adapted from Ref. [68]. Copyright 2021 Shiliang He et al.

of such metal ions. Willner et al. modified the termini of DNA origami nanostructures with consecutive G bases and controlled their assembly by the formation of G-quadruplexes at the interfaces (Fig. 1.13b) [110]. By the addition and removal of K^{I} ions, the nanostructures were reversibly dimerized and disassembled. Furthermore, Suzuki et al. applied G-quadruplexes for actuating a DNA origami construct (Fig. 1.13c) [111]. By introducing G-quadruplexes into the middle of DNA origami "nanoarms", the whole structure was curved only in the presence of K^{I} ions.

To expand the scope of input metal ions and target secondary structures, researchers have also been working on the introduction of artificial metal-binding sites into DNA assemblies. Sleiman et al. introduced synthetic ligands such as terpyridine (tpy) and diphenylphenanthroline (dpp) into the termini of DNA duplexes for metal-dependent thermal stabilization of the duplexes (Fig. 1.14a) [112]. It has been revealed that different metal ions were incorporated into the terminal metal-binding sites depending on the combination of the ligands. More recently, Xu et al. controlled SDRs using metal complexation at the termini of DNA duplexes (Fig. 1.14b) [113]. They introduced 8-hydroxyquinoline (8HQ) ligands into the termini of DNA strands as an independent "metal-responsive toehold". It was demonstrated that the interstrand 8HQ–Cu^{II}–8HQ complexes were formed by the addition of Cu^{II} ions, resulting in the progress of the SDRs. The metal-dependent structural conversion was possible not only for duplexes, but also for other assemblies. For instance, Clever et al. reported that complexation of pyridine (py) ligands and a Cu^{II} ion can be utilized for developing Cu^{II}-dependent G-quadruplexes [114–116]. The G-quadruplexes with terminal py–Cu^{II} complexes were applied for enantioselective catalysis (Fig. 1.14c) [116]. Shionoya et al. reported that DNA three-way-junctions (3WJs) modified with bipyridine (bpy) ligands can be thermally stabilized by 3:1 metal complexation at the junction core [117–121]. It has been successfully demonstrated that the structural conversion from DNA duplexes to 3WJs can be driven by the complexation of three bpy ligands and a Ni^{II} ion (Fig. 1.14d) [118]. These examples have demonstrated that metal-responsive DNA systems can be rationally designed by carefully selecting a DNA structural motif as a target and the ligand–metal combination.

As described above, various external stimuli-responsive systems have been developed. Among them, metal complexation has many advantages as the trigger for dynamic control of DNA assemblies as follows. (i) Coordination bonding is inherently reversible, enabling the iterative structural switching by the addition and removal of metal ions. (ii) The metal–ligand interactions are highly selective when the appropriate combination is selected. (iii) Coordination bonding is highly stable compared to hydrogen bonding of the nucleobase pairs, allowing for efficient switching of DNA assemblies. Given these unique properties, further elucidation of the metal-responsive control of DNA assemblies would be of great significance.

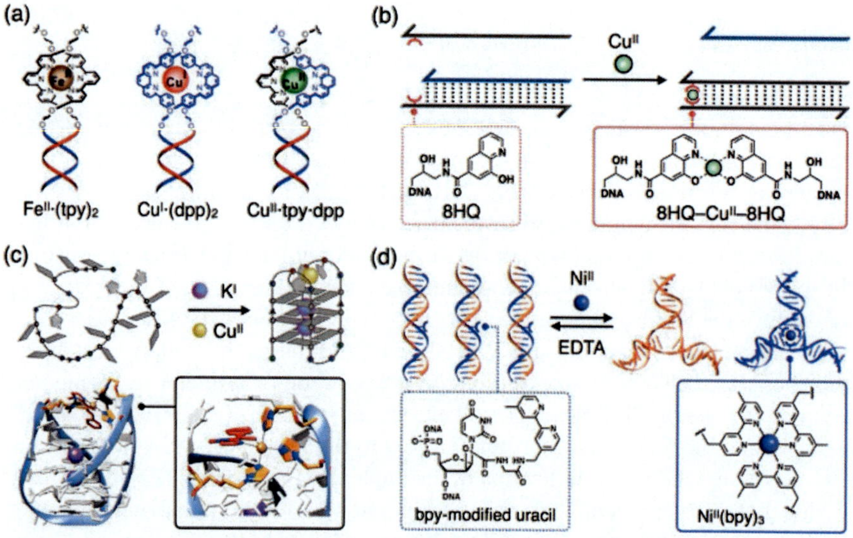

Fig. 1.14 Metal-responsive dynamic DNA systems using DNA strands with artificial metal-binding sites. **a** Thermal stabilization of DNA duplexes with terpyridine (tpy) and diphenylphenanthroline (dpp) ligands at the termini in the presence of different metal ions. Reprinted with permission from Ref. [77]. Copyright 2010 Elsevier. **b** Metal-responsive DNA strand displacement reactions (SDRs) driven by the formation of interstrand Cu^{II} complexes. **c** Cu^{II}-dependent formation of pyridine-modified G-quadruplexes and activation of catalysis. Adapted with permission from Ref. [116]. Copyright 2021 American Chemical Society. **d** Ni^{II}-dependent switching between DNA duplexes and three-way-junctions with $(bipyridine)^3$–Ni^{II} complexes at the core. Adapted from Ref. [121]. Copyright 2020 Yusuke Takezawa and Mitsuhiko Shionoya

Therefore, there is a need for a general strategy to make DNA-based structures metal-responsive.

1.4 Metal-Mediated Base Pairing

As mentioned in Sect. 1.3, metal complexation can be an excellent trigger for dynamic control of DNA assemblies. With a view to imparting metal responsiveness to DNA molecules, various strategies have been taken to introduce metal-binding sites into DNA strands. Shionoya et al. pioneered "metal-mediated base pairing" through 2:1 complexation of ligand-type nucleobases and a specific metal ion inside DNA duplexes (Fig. 1.15a) [104–107]. A pair of ligand-type nucleobases behave as a mismatch within DNA duplexes because there is no stabilizing interaction between them. When an appropriate metal ion is added, the two ligand-type nucleobases are bridged by coordination bonding, resulting in thermal stabilization of the duplexes. For example, hydroxypyridone (**H**) nucleoside was designed as a synthetic ligand-type nucleobase forming a metal-mediated base pair in the bidentate *O,O*-coordination mode (Fig. 1.15a) [122]. In fact, it was demonstrated that the metal-mediated **H**–Cu^{II}–**H** base pairs with a square-planar geometry can be constructed within DNA duplexes [123]. As a result, the addition of Cu^{II} ions significantly increased the melting temperature (T_m) of the duplex with an **H–H** mismatch ($\Delta T_m = + 13$ °C). Researchers have been exploiting various types of metal-mediated base pairs thus far. It has been well-known that natural thymine (T) bases form 2:1 complexes with a Hg^{II} ion through coordination of the N3 atom, resulting in T–Hg^{II}–T pairing (Fig. 1.15b) [102, 103]. Cytosine (C) bases were also found to show stable homo base pairing bridged by Ag^{I} ions (C–Ag^{I}–C pairs) [124]. The scope of bridging metal ions can be further expanded by introducing artificial nucleobase analogs into DNA strands (Fig. 1.15c). Schultz et al. developed a pyridine-2,6-dicaboxylate (**Dipic**) nucleotide exhibiting Cu^{II}-mediated hetero base pairing with a pyridine (**Py**) nucleotide [125]. Carell et al. reported that a salen (**S**)-bearing nucleotide forms metal-mediated base pairs with Mn^{II} or Cu^{II} ions with the aid of covalent crosslinking by ethylenediamine (en) [126].

The formation of the metal-mediated base pairs within DNA duplexes has been applicable to induce physical properties deriving from metal complexes. For instance, Shionoya et al. reported one-dimensional array of five Cu^{II} ions templated by DNA duplexes containing **H–H** pairs, resulting in ferromagnetic coupling among Cu^{II} ions (Fig. 1.16a) [127]. Carell et al. constructed two consecutive **S**–Cu^{II}(en)–**S** pairs inside duplexes, which exhibited antiferromagnetic coupling [128]. It was also demonstrated that two different metal-mediated base pairs can be incorporated into one DNA duplex, resulting in the heterometallic array [129]. Furthermore, Kondo et al. recently reported a crystal structure of a long uninterrupted Ag^{I} array templated by DNA duplexes [130]. Accordingly, it has attracted increasing attention to align metal ions inside DNA duplexes through metal-mediated base pairing, exploring novel properties of one-dimensional assembly of metal ions. The metal-mediated

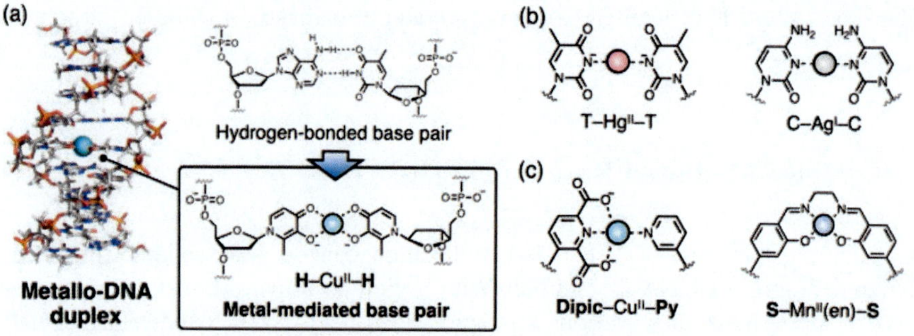

Fig. 1.15 Metal-mediated base pairing. **a** Concept of metal-mediated base pair. An **H–Cu^{II}–H** base pair is shown as an example. **b** Metal-mediated base pairs with natural nucleobases as ligands. **c** Metal-mediated base pairs using synthetic ligand-type nucleotides

base pairing was further utilized for switching physical properties. The conductivity of DNA duplexes was reversibly controlled depending on the formation of **H–Cu^{II}–H** base pairs (Fig. 1.16b) [131].

DNA assemblies can be dynamically controlled by the metal-mediated base pairing. Stable coordination bonding can efficiently drive the structural conversion of DNA assemblies. Ono et al. found that DNA single strands can be folded into hairpin structures by the formation of C–Ag^{I}–C base pairs (Fig. 1.17a) [124]. This metal-responsive hairpin assembly has been applicable to sensing devices targeting specific metal ions. Furthermore, Luedtke et al. demonstrated that T–Hg^{II}–C base pairing can induce dynamic transition between two types of DNA duplexes with different helical pitches, i.e., A- and B-form duplexes (Fig. 1.17b) [132]. Furthermore, the metal-responsive structural conversion was demonstrated using artificial ligand-type

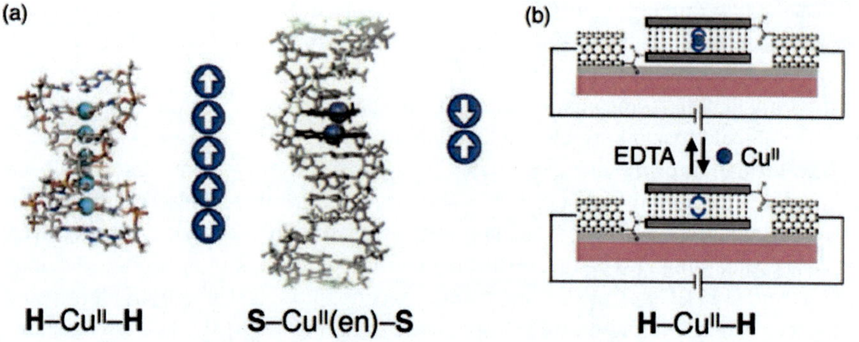

Fig. 1.16 Metal-mediated base pairs showing unique physical properties. **a** Cu^{II} arrays inside DNA duplexes showing magnetic properties. Adapted with permission from Ref. [105]. Copyright 2012 American Chemical Society. **b** Cu^{II}-dependent switching of the conductivity of DNA duplexes containing **H–Cu^{II}–H** base pairs. Adapted with permission from Ref. [105]. Copyright 2012 American Chemical Society

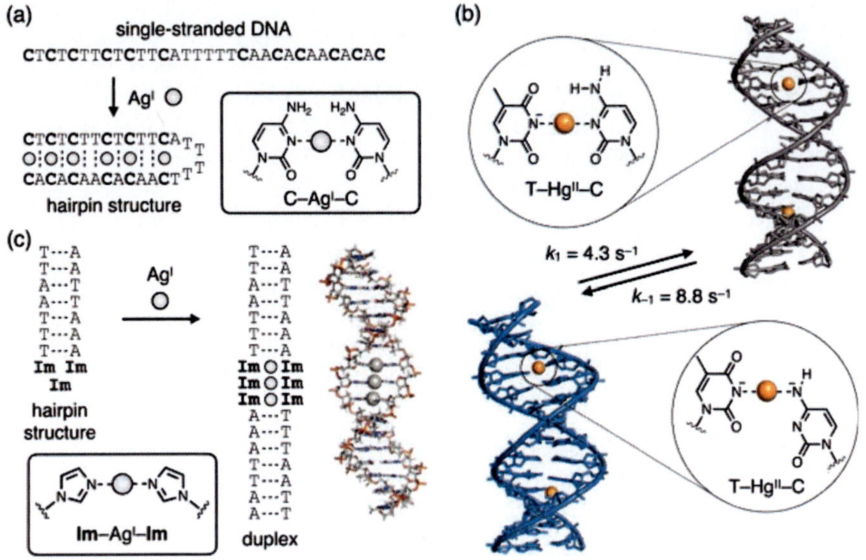

Fig. 1.17 Metal-dependent dynamic structural conversion of DNA. **a** Ag^I-dependent formation of DNA hairpins with C–Ag^I–C pairs. **b** Dynamic helical transition between A- and B-form DNA duplexes by T–Hg^{II}–C base pairing. Adapted with permission from Ref. [132]. Copyright 2019 Olivia P. Schmidt et al. **c** Ag^I-responsive transformation of DNA hairpins to duplexes through Ag^I-mediated base pairing of artificial nucleotides. Adapted with permission from Ref. [105]. Copyright 2012 American Chemical Society

nucleotides. Müller et al. reported that the formation of Ag^I-mediated base pairs triggers conformational changes of DNA from hairpins to duplexes (Fig. 1.17c) [133, 134]. Nucleotides bearing a metal ligand, 1,2,4-triazole or imidazole (**Im**), were introduced into the loop region of the hairpins. The homodimerization of the ligand-modified strands proceeded by the addition of Ag^I ions, yielding the Ag^I-mediated DNA duplexes.

The metal-mediated base pairing has been further applied to metal-dependent control of dynamic DNA nanodevices. Willner et al. developed a DNA molecular machine exhibiting a progressive motion, namely "DNA walker", based on Hg^{II}- and pH-dependent dynamic hybridization of DNA strands (Fig. 1.18a) [135]. In the initial state, two "feet" of the walker are fixed onto the first and the second footholds. Hg^{II} addition induced the T–Hg^{II}–T base pairing between one foot and the third foothold, resulting in the progressive step of the walker. pH-dependent formation of the i-motifs in the second foothold can induce the subsequent step. They further developed dynamic DNA catenanes based on similar mechanisms using the T–Hg^{II}–T base pairing and i-motif formation (Fig. 1.18b) [136]. Park et al. reported Hg^{II}-dependent control of enzymatic reactions (Fig. 1.18c) [137]. A DNA aptamer binding to a Taq DNA polymerase was used to inhibit the enzyme activity. When Hg^{II} ions were added, structural conversion of the aptamer sequence was induced by the formation of the intrastrand T–Hg^{II}–T pairs. As a result, the polymerase is released from the aptamer to recover its activity, exhibiting primer extension reactions. As described above, most of the metal-responsive DNA systems have relied on the T–Hg^{II}–T and C–Ag^{I}–C base pairing. Shionoya et al. recently reported that activity of RNA-cleaving DNAzymes can be controlled by metal complexation of synthetic **H** nucleotides (Fig. 1.18d) [138–140]. One **H–H** mismatch was introduced into the stem region of reported DNAzyme sequences. In the absence of Cu^{II} ions, the two strands cannot give stable duplexes due to the **H–H** mismatch. When Cu^{II} ions were added, they can be assembled by the formation of the **H–**Cu^{II}**–H** pair, resulting in regeneration of the catalytic core. As a result, the **H**-bearing DNAzyme showed enhanced activity upon addition of Cu^{II} ions. This strategy of metal-responsive allosteric DNAzymes has been further expanded to the DNAzymes showing logic gate responses [139] and using other Cu^{II}-binding nucleobases [141, 142].

Thus, it has been demonstrated that DNA duplexes can be thermally stabilized by the formation of metal-mediated base pairs. Based on this significant stabilization effects, the ligand-type nucleotides have been utilized as a metal-responsive molecular unit for dynamic control of DNA assemblies. However, these metal-responsive systems are still in its infancy because they mostly rely on T–Hg^{II}–T or C–Ag^{I}–C base pairing. Although several Cu^{II}-responsive systems have been developed using synthetic ligand-type nucleotides, the variety of the input metal ions is still limited. In addition, thermal stability of metal-free states is severely diminished due to the incorporation of ligand-bearing mismatch pairs. Therefore, it has been demanded to develop a new design of ligand-type nucleotides showing (i) flexible designability for targeting various metal ions as inputs and (ii) structural and chemical compatibility with natural DNA assemblies.

1.5 Bifacial Base-Pairing Behaviors of 5-Modified Uracil Bases

As described in Sect. 1.4, it has been proven that the metal-mediated base pairing has a great potential as a trigger of metal-responsive dynamic DNA systems. With a view to further expanding the applicability of the metal-mediated base pairing in dynamic DNA nanotechnology, there remain the following problems as the next challenges. (i) Scope of metal ions as the input is still limited. (ii) The introduction of ligand-type nucleotides affects the stability of DNA assemblies. To solve these problems, a 5-modified uracil base (**U**$^{\mathbf{X}}$) was designed as a "bifacial" nucleobase (Fig. 1.19). The **U**$^{\mathbf{X}}$ base possesses a metal-binding functionality (X) at the 5 position of a natural uracil, one of the canonical nucleobases. Since the **U**$^{\mathbf{X}}$ base retains the hydrogen-bonding site of uracil or thymine, it can form a hydrogen-bonded **U**$^{\mathbf{X}}$–A base pair with the adenine (A) base. On the other hand, the coordination of the 5-substituent (X) and 4-carbonyl groups to a metal ion would yield a metal-mediated **U**$^{\mathbf{X}}$–M–**U**$^{\mathbf{X}}$ base pair (M = metal ion). It has been expected that the "bifacial" **U**$^{\mathbf{X}}$ base has great

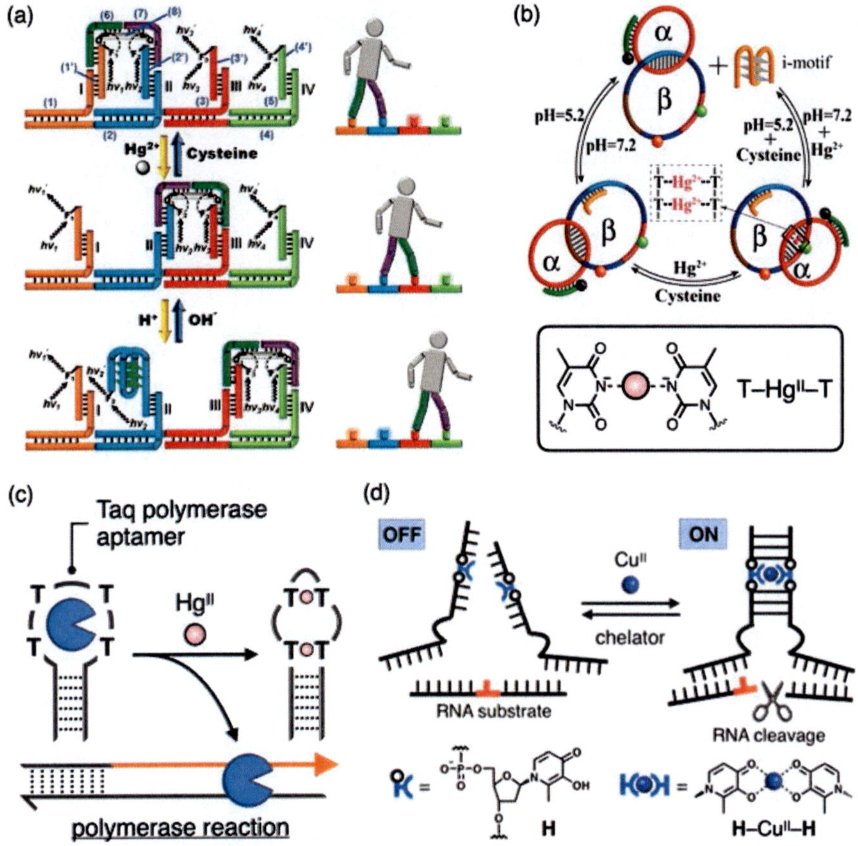

Fig. 1.18 Metal-dependent dynamic DNA nanodevices. **a** DNA walker operating in response to Hg^{II} ions and pH changes. Adapted with permission from Ref. [135]. Copyright 2011 American Chemical Society. **b** Hg^{II} and pH-responsive dynamic DNA catenanes. Adapted with permission from Ref. [136]. Copyright 2013 American Chemical Society. **c** Hg^{II}-triggered activation of Taq DNA polymerase. **d** Cu^{II}-dependent regulation of RNA-cleaving DNAzyme. Adapted with permission from Ref. [138]. Copyright 2019 American Chemical Society

potential as a molecular unit for metal-responsive dynamic DNA systems. Since the $\mathbf{U^X}$ base forms the $\mathbf{U^X}$–A pairs, the introduction of $\mathbf{U^X}$ will not greatly affect the structure and thermal stability of DNA assemblies. In addition, affinity toward metal ions can be tuned depending on the ligands (X) introduced into the 5 position.

Shionoya et al. have previously reported that the 5-hydroxyuracil ($\mathbf{U^{OH}}$) nucleobases show Gd^{III}-dependent bifacial base-pairing behaviors (Fig. 1.20a) [143–145]. Since the $\mathbf{U^{OH}}$ base retains the hydrogen-bonding site, it can form hydrogen-bonded $\mathbf{U^{OH}}$–A base pairs. It was also revealed that the $\mathbf{U^{OH}}$ base behaves as a bidentate ligand through the coordination of the 4-carbonyl and 5-hydroxy groups in the *O*,*O*-binding mode, allowing for the formation of metal-mediated $\mathbf{U^{OH}}$–Gd^{III}–$\mathbf{U^{OH}}$ pairs. The base-pairing properties of $\mathbf{U^{OH}}$ were studied by duplex melting analysis using

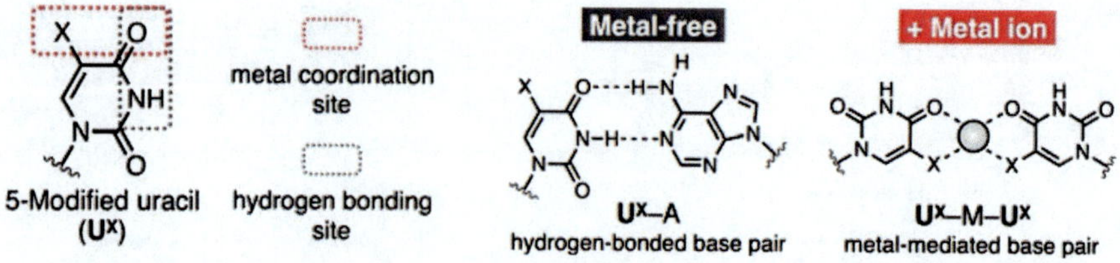

Fig. 1.19 Concept of metal-dependent bifacial base pairing of 5-modified uracil (**U**$^{\mathbf{X}}$) bases

DNA duplexes containing three **U**$^{\mathbf{OH}}$–A or **U**$^{\mathbf{OH}}$–**U**$^{\mathbf{OH}}$ pairs (Fig. 1.20b). Under Gd^{III}-free conditions, the duplex with **U**$^{\mathbf{OH}}$–A pairs showed higher stability (blue dotted line, $T_m = 45.2$ °C) than the other duplex with **U**$^{\mathbf{OH}}$–**U**$^{\mathbf{OH}}$ pairs (red dotted line, $T_m = 22.8$ °C). When 3 equiv. of Gd^{III} ions were added, the duplex with **U**$^{\mathbf{OH}}$–**U**$^{\mathbf{OH}}$ pairs was greatly stabilized (red solid line, $T_m = 41.1$ °C, $\Delta T_m = +\ 18.3$ °C) by the formation of metal-mediated **U**$^{\mathbf{OH}}$–Gd^{III}–**U**$^{\mathbf{OH}}$ pairs. In contrast, thermal stability of the other duplex with **U**$^{\mathbf{OH}}$–A pairs was lowered (blue solid line, $T_m = 30.9$ °C, $\Delta T_m = -\ 14.3$ °C) possibly because the hydrogen bonding between **U**$^{\mathbf{OH}}$ and A was weakened by the binding of Gd^{III} ions to the **U**$^{\mathbf{OH}}$ bases. As a result, relative stability of the two duplexes was inverted by the addition of Gd^{III} ions. This result indicates that the preferential base-pairing partner of the **U**$^{\mathbf{OH}}$ base can be altered in a Gd^{III}-dependent manner.

The metal selectivity of the **U**$^{\mathbf{OH}}$ base was also investigated. It was revealed that the duplex with **U**$^{\mathbf{OH}}$–**U**$^{\mathbf{OH}}$ pairs is stabilized in the presence of other lanthanide ions such as Eu^{III} and Tb^{III} as well ($\Delta T_m = +\ 14.0$–20.3 °C). On the other hand, Zn^{II} ions showed only moderate stabilizing effects ($\Delta T_m = +\ 13.0$ °C) and other transition metal ions such as Cu^{II} and Mn^{II} did not stabilize the duplex at all. These results indicate that the **U**$^{\mathbf{OH}}$ bases exhibit high selectivity toward lanthanide ions.

A carboxy group was also introduced at the 5 position of the **U**$^{\mathbf{X}}$ base as a metal-binding ligand, and a 5-carboxyuracil (**caU**) nucleobase was designed (Fig. 1.21) [146]. It was revealed that the **caU** base shows Cu^{II}-mediated **caU**–Cu^{II}–**caU** base pairing within DNA duplexes while it can also form hydrogen-bonded **caU**–A pairs. Accordingly, it has been demonstrated that the metal selectivity of the 5-modified uracil **U**$^{\mathbf{X}}$ can be altered by changing the ligand at the 5 position.

Thus, the 5-modified uracil bases form hydrogen-bonded base pairs and metal-mediated base pairs in the absence and presence of specific metal ions, respectively. This bifacial base-pairing behavior allows for stabilization of two different assembly states with different base-pairing modes. It is expected that the addition of input metal ions can efficiently drive the structural conversion from metal-free to metal-containing assemblies. Therefore, the 5-modified uracil bases would be useful as a molecular tool to construct metal-responsive dynamic DNA supramolecules.

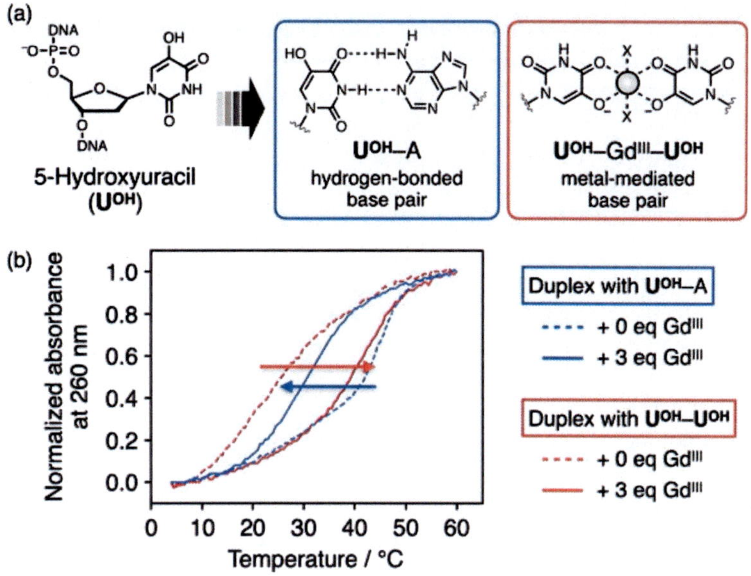

Fig. 1.20 Gd^{III}-dependent bifacial base-pairing behaviors of 5-hydroxyuracil (**U^{OH}**) bases. **a** Chemical structures of the hydrogen-bonded **U^{OH}**–A pair and the metal-mediated **U^{OH}**–Gd^{III}–**U^{OH}** pair. "X" denotes additional ligands coordinating to the Gd^{III} ion such as H_2O and neighboring nucleobases. **b** Melting curves of DNA duplexes containing three **U^{OH}**–A or **U^{OH}**–**U^{OH}** pairs in the absence and presence of Gd^{III} ions. The inflection point of the curve denotes a melting temperature (T_m), representing the thermal stability of the duplex

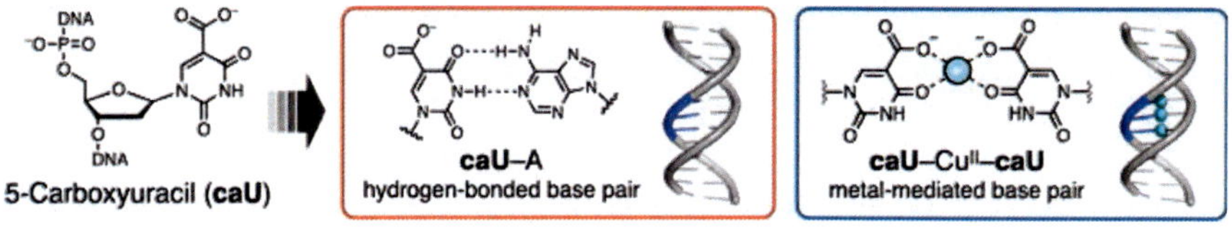

Fig. 1.21 Cu^{II}-dependent bifacial base-pairing behaviors of 5-carboxyuracil (**caU**). Adapted with permission from Ref. [146]. Copyright 2020 American Chemical Society

1.6 The Aim of This Study

As discussed above, dynamic control of DNA assemblies in response to external stimuli is important to expand the scope of dynamic DNA nanotechnology. In particular, metal complexation has been recognized as a useful trigger due to its reversibility and ligand specificity. Although metal-mediated base pairing has emerged as a tool for fabricating DNA-based supramolecules that respond to metal ions, the applicability to dynamic DNA nanotechnology is still limited. Therefore, there is a need to establish a general strategy for the design of ligand-based nucleobases suitable for metal-responsive dynamic conversion of DNA structures. In this study, I aimed to dynamically and reversibly control DNA assemblies based on the metal-dependent bifacial base pairing of 5-modified uracil (**U^{X}**) bases. For this purpose, I proposed a concept of metal-responsive "base pair switching" of the 5-modified uracil. As described in Sect. 1.5, the **U^{X}** base can form both hydrogen-bonded base pairs and metal-mediated base pairs depending on the presence of specific metal ions. Accordingly, it has been expected that the interconversion of the two different base-pairing modes, named base pair switching, can drive dynamic transformation of DNA structures.

As an approach to metal-responsive dynamic DNA systems, I focused on DNA strand displacement reactions (SDRs), which are an essential tool in dynamic DNA nanotechnology. I demonstrated the metal-responsive base pair switching of 5-hydroxyuracil (**U^{OH}**) nucleobases (Fig. 1.22a) and applied it to the control of SDRs (Fig. 1.22b). It was confirmed that the base-pairing partner of **U^{OH}** can be switched by the addition of Gd^{III} ions. Furthermore, the SDR was successfully regulated based on the Gd^{III}-responsive base pair switching of **U^{OH}**.

Despite this successful demonstration of SDRs, it was known that the consecutive incorporation of **U^{OH}** was required for **U^{OH}**–Gd^{III}–**U^{OH}** complexation. This limitation greatly reduces the degree of flexibility in the sequence design. Therefore, I aimed to develop a new bifacial nucleobase that forms a single metal-mediated base pair inside DNA duplexes. For this purpose, *N*,*N*-dicarboxymethyl-5-aminouracil (**dcaU**) nucleoside possessing an iminodiacetic acid ligand at the 5 position was newly synthesized and incorporated into DNA strands (Fig. 1.22c). In fact, the **dcaU** base showed Gd^{III}-dependent base pair switching just by the single incorporation into DNA strands.

Finally, I aimed to show that metal-responsive base pair switching is generally applicable to dynamic DNA nanotechnology. Accordingly, it was applied to dynamic regulation of DNA nanodevices (Fig. 1.22d). DNA tweezers, a well-known simple DNA molecular machine, were modified with the **U^{OH}** bases. It was confirmed that the tweezers can be reversibly operated by the metal-mediated base pair switching under isothermal conditions.

In summary, the purpose of this study is to establish the "base pair switching" of 5-modified uracil bases as a novel molecular tool for the development of metal-responsive DNA-based dynamic systems. I believe that this study proves the applicability of the metal-responsive base pair switching and broaden the scope of dynamic DNA nanotechnology.

Chapter 2 describes the development of Gd^{III}-triggered SDRs based on the base pair switching of **U^{OH}** (Fig. 1.22a, b). Gd^{III}-responsive reversible SDRs were demonstrated by introducing four **U^{OH}** bases at the termini of DNA strands as Gd^{III}-binding sites.

Chapter 3 demonstrates the Gd^{III}-dependent base pair switching of the **dcaU** base (Fig. 1.22c). The **dcaU** nucleoside was newly synthesized and incorporated into DNA strands. The Gd^{III}-dependent base pair switching was confirmed with the single incorporation of **dcaU**.

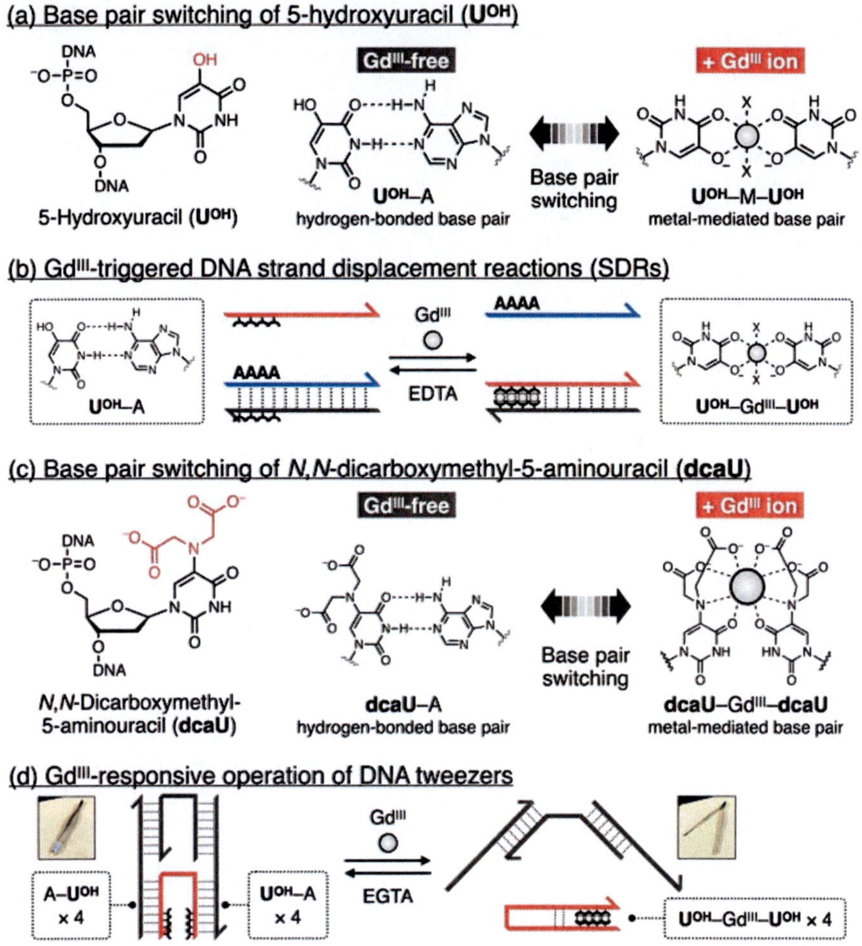

Fig. 1.22 Outline of this study

Chapter 4 illustrates Gd^{III}-responsive control of DNA tweezers (Fig. 1.22d). By incorporating multiple **U^{OH}** bases into the arm of the DNA tweezers, they were reversibly operated in response to Gd^{III} ions under isothermal conditions.

References

1. Zuker M (2003) Nucleic Acids Res 31:3406–3415
2. Zadeh JN, Steenberg CD, Bois JS, Wolfe BR, Pierce MB, Khan AR, Dirks RM, Pierce NA (2011) J Comput Chem 32:170–173
3. Seeman NC, Sleiman HF (2017) Nat Rev Mater 3:17068
4. Seeman NC (2016) Structural DNA nanotechnology. Cambridge University Press, Cambridge

5. Zhang F, Nangreave J, Liu Y, Yan H (2014) J Am Chem Soc 136:11198–11211
6. Hu Q, Li H, Wang L, Gu H, Fan C (2019) Chem Rev 119:6459–6506
7. Xiao M, Lai W, Man T, Chang B, Li L, Chandrasekaran AR, Pei H (2019) Chem Rev 119:11631–11717
8. Seeman NC (1982) J Theor Biol 99:237–247
9. Kallenbach NR, Ma R-I, Seeman NC (1983) Nature 305:829–831
10. Chen J, Seeman NC (1991) Nature 350:631–633
11. Seeman NC (2003) Nature 421:427–431
12. Goodman RP, Berry RM, Turberfield AJ (2004) Chem Commun 1372–1373
13. Goodman RP, Schaap IAT, Tardin CF, Erben CM, Berry RM, Schmidt CF, Turberfield AJ (2005) Science 310:1661–1665
14. Fu T-J, Seeman NC (1993) Biochemistry 32:3211–3220
15. Winfree E, Liu F, Wenzler LA, Seeman NC (1998) Nature 394:539–544
16. Lin C, Liu Y, Rinker S, Yan H (2006) ChemPhysChem 7:1641–1647
17. Yan H, Park SH, Finkelstein G, Reif JH, LaBean TH (2003) Science 301:1882–1884
18. He Y, Chen Y, Liu H, Ribbe AE, Mao C (2005) J Am Chem Soc 127:12202–12203
19. Sun X, Ko SH, Zhang C, Ribbe AE, Mao C (2009) J Am Chem Soc 131:13248–13249
20. He Y, Ye T, Su M, Zhang C, Ribbe AE, Jiang W, Mao C (2008) Nature 452:198–202
21. Zheng J, Birktoft JJ, Chen Y, Wang T, Sha R, Constantinou PE, Ginell SL, Mao C, Seeman NC (2009) Nature 461:74–77
22. Rothemund PWK (2006) Nature 440:297–302
23. Hong F, Zhang F, Liu Y, Yan H (2017) Chem Rev 117:12584–12640
24. Tørring T, Voigt NV, Nangreave J, Yan H, Gothelf KV (2011) Chem Soc Rev 40:5636–5646
25. Andersen ES, Dong M, Nielsen MM, Jahn K, Subramani R, Mamdouh W, Golas MM, Sander B, Stark H, Oliveira CLP, Pedersen JS, Birkedal V, Besenbacher F, Gothelf KV, Kjems J (2009) Nature 459:73–77
26. Han D, Pal S, Nangreave J, Deng Z, Liu Y, Yan H (2011) Science 332:342–346
27. Douglas SM, Dietz H, Liedl T, Högberg B, Graf F, Shih WM (2009) Nature 459:414–418
28. Ke Y, Voigt NV, Gothelf KV, Shih WM (2012) J Am Chem Soc 134:1770–1774
29. Gerling T, Wagenbauer TF, Neuner AM, Dietz H (2015) Science 347:1446–1452
30. Nummelin S, Shen B, Piskunen P, Liu Q, Kostiainen MA, Linko V (2020) ACS Synth Biol 9:1923–1940
31. Ji J, Karna D, Mao H (2021) Chem Soc Rev 50:11966–11978
32. Ramezani H, Dietz H (2020) Nat Rev Genet 21:5–26
33. Kuzuya A, Ohya Y (2014) Acc Chem Res 47:1742–1749
34. Douglas SM, Marblestone AH, Teerapittayanon S, Vazquez A, Church GM, Shih WM (2009) Nucleic Acids Res 37:5001–5006
35. Madsen M, Gothelf KV (2019) Chem Rev 119:6384–6458
36. Voigt NV, Tørring T, Rotaru A, Jacobsen MF, Ravansbæk JB, Subramani R, Mamdouh W, Kjems J, Mokhir A, Besenbacher F, Gothelf KV (2010) Nat Nanotechnol 5:200–203
37. Fu J, Liu M, Liu Y, Woodbury NW, Yan H (2012) J Am Chem Soc 134:5516–5519
38. Dutta PK, Varghese R, Nangreave J, Lin S, Yan H, Liu Y (2011) J Am Chem Soc 133:11985–11993
39. Langecker M, Arnaut V, Martin TG, List J, Renner S, Mayer M, Dietz H, Simmel FC (2012) Science 338:932–936
40. Zhang DY, Seelig G (2011) Nat Chem 3:103–113
41. DeLuca M, Shi Z, Castro CE, Arya G (2020) Nanoscale Horiz. 5:182
42. Simmel FC, Yurke B, Singh HR (2019) Chem Rev 119:6326–6369
43. Yurke B, Mills AP (2003) Genet Prog Evol Mach 4:111
44. Zhang DY, Winfree E (2009) J Am Chem Soc 131:17303–17314
45. Mao X, Liu M, Li Q, Fan C, Zuo X (2022) JACS Au 2:2381–2399
46. Guo Y, Wei B, Xiao S, Yao D, Li H, Xu H, Song T, Li X, Liang H (2017) Quant Biol 5:25–41
47. Yurke B, Turberfield AJ, Mills AP, Simmel FC, Neumann JL (2000) Nature 406:605–608
48. Shin JS, Pierce NA (2004) J Am Chem Soc 126:10834–10835

49. Yin P, Yan H, Daniell XG, Turberfield AJ, Reif JH (2004) Angew Chem Int Ed 43:4906–4911
50. Tian Y, Mao C (2004) J Am Chem Soc 126:11410–11411
51. Grossi G, Jepsen MDE, Kjems J, Andersen ES (2017) Nat Commun 8:992
52. Han D, Pal S, Lu Y, Yan H (2010) Nat Nanotechnol 5:712–717
53. Dirks RM, Pierce NA (2004) Proc Natl Acad Sci USA 101:15275–15278
54. Bi S, Yue S, Zhang S (2017) Chem Soc Rev 46:4281–4298
55. Augspurger EE, Rana M, Yigit MV (2018) ACS Sens 3:878–902
56. Huang J, Wu Y, Chen Y, Zhu Z, Yang X, Yang CJ, Wang K, Tan W (2011) Angew Chem Int Ed 50:401–404
57. Song C, Li B, Yang X, Wang K, Wang Q, Liu J, Huang J (2017) Analyst 142:224–228
58. Cangialosi A, Yoon C, Liu J, Huang Q, Guo J, Nguyen TD, Gracias DH, Schulman R (2017) Science 357:1126–1130
59. Fern J, Schulman R (2018) Nat Commun 9:3766
60. Zhang DY, Turberfield AJ, Yurke B, Winfree E (2007) Science 318:1121–1125
61. Seelig G, Soloveichik D, Zhang DY, Winfree E (2006) Science 314:1585–1588
62. Lv H, Li Q, Shi J, Fan C, Wang F (2021) ChemPhysChem 22:1151–1166
63. Zhao S, Yu L, Yang S, Tang X, Chang K, Chen M (2021) Nanoscale Horiz 6:298–310
64. Qian L, Winfree E (2011) Science 332:1196–1201
65. Qian L, Winfree E, Bruck J (2011) Nature 475:368–372
66. Bucci J, Irmisch P, Grosso ED, Seidel R, Ricci F (2023) J Am Chem Soc 145:20968–20974
67. Lu S, Shen J, Fan C, Li Q, Yang X (2021) Adv Sci 8:2100328
68. He S, Ge Z, Zuo X, Fan C, Mao X (2021) NPG Asia Mater 13:42
69. Wang C, O'Hagan MP, Li Z, Zhang J, Ma X, Tian H, Willner I (2022) Chem Soc Rev 51:720–760
70. Lubbe AS, Szymanski W, Feringa BL (2017) Chem Soc Rev 46:1052–1079
71. Kamiya Y, Asanuma H (2014) Acc Chem Res 47:1663–1672
72. Mattiath MN, Ghosh D, Pratihar S, Shi S, Govindaraju T (2022) ACS Omega 7:367–3176
73. Krissanaprasit A, Key CM, Pontula S, LaBean TH (2021) Chem Rev 121:13797–13868
74. McConnell AJ, Wood CS, Neelakandan PP, Nitschke JR (2015) Chem Rev 115:7729–7793
75. Goswami A, Saha S, Biswas PK, Schmittel M (2020) Chem Rev 120:125–199
76. Chakrabarty R, Mukherjee PS, Stang PJ (2011) Chem Rev 111:6810–6918
77. Yang H, Metera KL, Sleiman HF (2010) Coord Chem Rev 254:2403–2415
78. Biver T (2013) Coord Chem Rev 257:2765–2783
79. Ihara T, Kitamura Y, Katsuda Y (2022) Life 12:686
80. Morihiro K, Kodama T, Waki R, Obika S (2014) Chem Sci 5:744–750
81. Li J, Wang X, Liang X (2014) Chem Asian J 9:3344–3358
82. Asanuma H, Ito T, Yoshida T, Liang X, Komiyama M (1999) Angew Chem Int Ed 38:2393–2395
83. Kou B, Zhang J, Huai X, Liang X, Xiao SJ (2015) RSC Adv 5:5055–5058
84. Cheng B, Kashida H, Shimada N, Maruyama A, Asanuma H (2017) ChemBioChem 18:1568–1572
85. Willner EM, Kamada Y, Suzuki Y, Emura T, Hidaka K, Dietz H, Sugiyama H, Endo M (2017) Angew Chem Int Ed 129:15526–15530
86. Lohmann F, Ackermann D, Famulok M (2012) J Am Chem Soc 134:11884–11887
87. Jester S-S, Famulok M (2014) Acc Chem Res 47:1700–1709
88. Sinha RP, Häder D-P (2002) Photochem Photobiol Sci 1:225–236
89. Bléger D, Hecht S (2015) Angew Chem Int Ed 54:11338–11349
90. Kamiya Y, Arimura Y, Ooi H, Kato K, Liang X-G, Asanuma H (2018) ChemBioChem 19:1305–1311
91. Dawson RMC, Elliott DC, Elliott WH, Jones KM (1986) Data for biochemical research, 3rd edn. Oxford University Press, Oxford
92. Amodio A, Zhao B, Porchetta A, Idili A, Castronovo M, Fan C, Ricci F (2014) J Am Chem Soc 136:16469–16472
93. Wu N, Willner I (2016) Nano Lett 16:6650–6655

94. Assi HA, Garavís M, González C, Damha MJ (2018) Nucleic Acids Res 46:8038–8056
95. Kim SH, Kim K-R, Ahn D-R, Lee JE, Yang EG, Kim SY (2017) ACS Nano 11:9352–9359
96. Guo W, Lu C-H, Orbach R, Wang F, Qi X-J, Cecconello A, Seliktar D, Willner I (2015) Adv Mater 27:73–78
97. Kahn JS, Hu Y, Willner I (2017) Acc Chem Res 50:680–690
98. Eaton BE (1997) Curr Opin Chem Biol 1:10–16
99. Ni S, Zhuo Z, Pan Y, Yu Y, Li F, Liu J, Wang L, Wu X, Li D, Wan Y, Zhang L, Yang Z, Zhang B-T, Lu A, Zhang G (2021) ACS Appl Mater Interfaces 13:9500–9510
100. Zhang Q-L, Wang L-L, Liu Y, Lin J, Xu L (2021) Nat Commun 12:4654
101. Kuzuya A, Sakai Y, Yamazaki T, Xu Y, Komiyama M (2011) Nat Commun 2:449
102. Katz S (1963) Biochim Biophys Acta 68:240–253
103. Miyake Y, Togashi H, Tashiro M, Yamaguchi H, Oda S, Kudo M, Tanaka Y, Kondo Y, Sawa R, Fujimoto T, Machinami T, Ono A (2006) J Am Chem Soc 128:2172–2173
104. Takezawa Y, Müller J, Shionoya M (2017) Chem Lett 46:622–633
105. Takezawa Y, Shionoya M (2012) Acc Chem Res 45:2066–2076
106. Müller J (2019) Coord Chem Rev 393:37–47
107. Naskar S, Guha R, Müller J (2020) Angew Chem Int Ed 59:1397–1406
108. Ding W, Deng W, Zhu H, Liang H (2013) Chem Commun 49:9953–9955
109. Dong J, O'Hagan MP, Willner I (2022) Chem Soc Rev 51:7631–7661
110. Wang J, Yue L, Wang S, Willner I (2018) ACS Nano 12:12324–12336
111. Suzuki Y, Kawamata I, Mizuno K, Murata S (2020) Angew Chem Int Ed 59:6230–6234
112. Yang H, Rys AZ, McLaughlin CK, Sleiman HF (2009) Angew Chem Int Ed 48:9919–9923
113. Wang L-L, Zhang Q-L, Wang Y, Liu Y, Lin J, Xie F, Xu L (2021) Chem Sci 12:8698–8705
114. Engelhard DM, Pievo R, Clever GH (2013) Angew Chem Int Ed 52:12843–12847
115. Engelhard DM, Nowack J, Clever GH (2017) Angew Chem Int Ed 56:11640–11644
116. Punt PM, Langenberg MD, Altan O, Clever GH (2021) J Am Chem Soc 143:3555–3561
117. Duprey J-LHA, Takezawa Y, Shionoya M (2013) Angew Chem Int Ed 52:1212–1216
118. Takezawa Y, Yoneda S, Duprey J-LHA, Nakama T, Shionoya M (2016) Chem Sci 7:3006–3010
119. Takezawa Y, Sakakibara S, Shionoya M (2021) Chem Eur J 67:16626–16633
120. Takezawa Y, Kanemaru D, Kudo N, Shionoya M (2023) Dalton Trans 52:11025–11029
121. Takezawa Y, Shionoya M (2020) Front Chem 7:925
122. Tanaka K, Shionoya M (1999) J Org Chem 64:5002–5003
123. Tanaka K, Tengeiji A, Kato T, Toyama N, Shiro M, Shionoya M (2002) J Am Chem Soc 124:12494–12498
124. Ono A, Cao S, Togashi H, Tashiro M, Fujimoto T, Machinami T, Oda S, Miyake Y, Okamoto I, Tanaka Y (2008) Chem Commun 44:4825–4827
125. Meggers E, Holland PL, Tolman WB, Romesberg FE, Schultz PG (2000) J Am Chem Soc 122:10714–10715
126. Clever GH, Polborn K, Carell T (2005) Angew Chem Int Ed 44:7204–7208
127. Tanaka K, Tengeiji A, Kato T, Toyama N, Shionoya M (2003) Science 299:1212–1213
128. Clever GH, Reitmeier SJ, Carell T, Schiemann O (2010) Angew Chem Int Ed 49:4927–4929
129. Tanaka K, Clever GH, Takezawa Y, Yamada Y, Kaul C, Shionoya M, Carell T (2006) Nat Nanotechnol 1:190–194
130. Kondo J, Tada Y, Dairaku T, Hattori Y, Saneyoshi H, Ono A, Tanaka Y (2017) Nat Chem 9:956–960
131. Liu S, Clever GH, Takezawa Y, Kaneko M, Tanaka K, Guo X, Shionoya M (2011) Angew Chem Int Ed 50:8886–8890
132. Schmidt OP, Jurt S, Johannsen S, Karimi A, Sigel RKO, Luedtke NW (2019) Nat Commun 10:4818
133. Böhme D, Düpre N, Megger DA, Müller J (2007) Inorg Chem 46:10114–10119
134. Johannsen S, Megger N, Böhme D, Sigel RKO, Müller J (2010) Nat Chem 2:229–234
135. Wang Z-G, Elbaz J, Willner I (2011) Nano Lett 11:304–309
136. Lu C-H, Cecconello A, Elbaz J, Credi A, Willner I (2013) Nano Lett 13:2303–2308
137. Park KS, Lee CY, Park HG (2016) Chem Commun 52:4868–4871

138. Takezawa Y, Nakama T, Shionoya M (2019) J Am Chem Soc 141:19342–19350
139. Nakama T, Takezawa Y, Sasaki D, Shionoya M (2020) J Am Chem Soc 142:10153–10162
140. Nakama T, Takezawa Y, Shionoya M (2021) Chem Commun 57:1392–1395
141. Takezawa Y, Hu L, Nakama T, Shionoya M (2020) Angew Chem Int Ed 59:21488–21492
142. Rajasree SC, Takezawa Y, Shionoya M (2023) Chem Commun 59:1006–1009
143. Takezawa Y, Nishiyama K, Mashima T, Katahira M, Shionoya M (2015) Chem Eur J 21:14713–14716
144. Nishiyama K, Takezawa Y, Shionoya M (2016) Inorg Chim Acta 452:176–180
145. Nishiyama K, Mori K, Takezawa Y, Shionoya M (2021) Chem Commun 57:2487–2490
146. Takezawa Y, Suzuki A, Nakaya M, Nishiyama K, Shionoya M (2020) J Am Chem Soc 142:21640–21644

Chapter 2
Metal-Responsive DNA Strand Displacement Reactions Driven by Base Pair Switching of 5-Hydroxyuracil Nucleobases

Abstract Toehold-mediated DNA strand displacement reactions (SDRs) are an essential tool for the construction of DNA-based molecular machines. In this chapter, I demonstrated "base pair switching" of 5-hydroxyuracil ($\mathbf{U^{OH}}$) bases and applied it for metal-responsive SDRs. The $\mathbf{U^{OH}}$ bases were known to form both hydrogen-bonded $\mathbf{U^{OH}}$–A pairs and metal-mediated $\mathbf{U^{OH}}$–Gd^{III}–$\mathbf{U^{OH}}$ pairs. Therefore, it was expected that the base-pairing partner of $\mathbf{U^{OH}}$ can be switched from A to $\mathbf{U^{OH}}$ by the addition of Gd^{III} ions, resulting in metal-responsive base pair switching. UV melting analysis revealed that the order of thermal stability was inverted among two $\mathbf{U^{OH}}$-containing duplexes, one with $\mathbf{U^{OH}}$–A pairs and the other with $\mathbf{U^{OH}}$–$\mathbf{U^{OH}}$ mismatches, by the addition of Gd^{III} ions. This result supported Gd^{III}-responsive base pair switching from $\mathbf{U^{OH}}$–A to $\mathbf{U^{OH}}$–Gd^{III}–$\mathbf{U^{OH}}$, which was utilized for metal-responsive SDRs. According to time-course fluorescence measurement, the addition of Gd^{III} ions triggered the SDR from the duplex containing $\mathbf{U^{OH}}$–A pairs to that with $\mathbf{U^{OH}}$–Gd^{III}–$\mathbf{U^{OH}}$ pairs. The SDR driven by the base pair switching of $\mathbf{U^{OH}}$ will be a powerful tool for developing metal-responsive DNA nanodevices.

Keywords DNA strand displacement reaction · Base pair switching · Metal-mediated base pairing · 5-hydroxyuracil · Metal complexation

2.1 Introduction

DNA strand displacement reactions (SDRs) have been widely utilized to develop dynamic DNA-based systems [1–3]. SDRs are normally mediated by hybridization of a terminal single-stranded region, termed as "toehold" (Fig. 2.1a). After the binding of an input strand to the toehold, "branch migration" proceeds to displace an output strand from the original duplex. As a result, another duplex with higher thermal stability is formed to complete the SDR. SDRs are highly programmable because they respond selectively to specific oligonucleotides and the reaction rate can be controlled depending on the toehold length [4]. Cascade SDRs are used to build complicated reaction networks such as logic circuits and oscillators [5–7]. Recently,

K. Mori, *Metal-Responsive Base Pair Switching of Ligand-type Uracil Nucleobases*, Springer Theses, https://doi.org/10.1007/978-981-99-9400-7_2

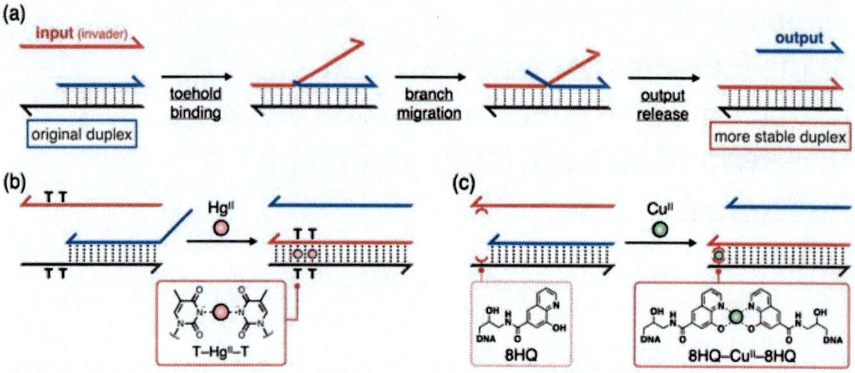

Fig. 2.1 Previous studies on metal-responsive strand displacement reactions (SDRs). **a** Conventional toehold-mediated SDRs using natural DNA strands. **b** Hg^{II}-responsive SDRs driven by toehold binding through the formation of T–Hg^{II}–T base pairs. **c** Cu^{II}-responsive SDRs triggered by 8HQ–Cu^{II}–8HQ complexation

SDRs triggered by external stimuli such as light irradiation [8–10], pH changes [11], and host–guest interactions [12] instead of oligonucleotides binding have gathered increasing attention to expand their applicability in DNA nanotechnology. These systems have been widely applied to stimuli-responsive DNA nanodevices [13–15] such as DNA-based molecular machines [16–18], controllable nanostructure assemblies [19, 20], and dynamic hydrogels [21, 22]. Among various external stimuli, I have focused on metal ions as a trigger to initiate SDRs because metal complexation is generally reversible and selective for specific ligands.

Metal-triggered SDRs have been previously demonstrated by introducing multiple T–T mismatches in a toehold region (Fig. 2.1b). SDRs proceeded only in the presence of Hg^{II} ions through the formation of metal-mediated T–Hg^{II}–T base pairs [23]. However, Hg^{II} ions can interact with any T bases of DNA strands coexisting in the solution. Therefore, it is still difficult to apply the Hg^{II}-responsive SDRs for controlling higher-ordered DNA nanostructures composed of many DNA strands. Accordingly, a more general strategy to rationally design SDRs triggered by specific metal ions has been demanded. Xu et al. have recently developed Cu^{II}-responsive SDRs by introducing 8-hydroxyquinoline (8HQ) ligands into the termini of DNA strands (Fig. 2.1c) [24]. Because the 8HQ ligand forms stable 2:1 complexes with Cu^{II} ions, SDRs can be initiated by interstrand 8HQ–Cu^{II}–8HQ complexation instead of the toehold binding in the traditional SDRs. Although this SDR showed a good selectivity for Cu^{II} ions, it required the introduction of an all-synthetic ligand with an unnatural backbone. This artificial modification would have a non-negligible effect on the stability and structure of the DNA assembly.

In this study, I aimed to develop metal-responsive SDRs using 5-hydroxyuracil (**U^OH^**) nucleobases (Fig. 2.2). As mentioned in Sect. 1.5, the **U^OH^** base behaves as a bidentate ligand with an *O,O*-binding mode through the coordination

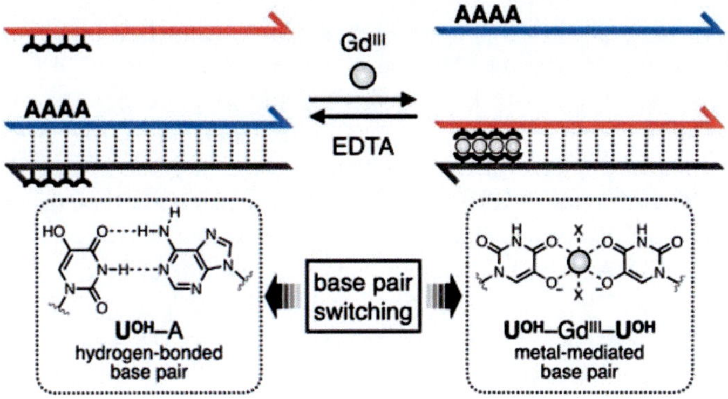

Fig. 2.2 Schematic illustration of a metal-responsive strand displacement reaction (SDR) driven by base pair switching of 5-hydroxyuracil (**U^{OH}**) bases. "X" denotes additional ligands coordinating to the Gd^{III} ion such as H_2O and neighboring nucleobases

of the 5-hydroxy and 4-carbonyl groups [25, 26]. Therefore, **U^{OH}** bases form metal-mediated **U^{OH}**–Gd^{III}–**U^{OH}** base pairs within DNA duplexes in the presence of Gd^{III} ions. Since the Watson–Crick faces of natural uracil and thymine bases are retained, **U^{OH}** bases also form hydrogen-bonded **U^{OH}**–A base pairs. In this study, metal-dependent interconversion of the two base-pairing modes, termed as "base pair switching", was proposed. It was expected that the Gd^{III}-responsive base pair switching of **U^{OH}** at the termini of DNA strands would efficiently drive SDRs in a manner similar to the canonical toehold-mediated SDRs. Since the **U^{OH}** bases can be easily incorporated into DNA strands and behaves as an analog of the natural T base through the **U^{OH}**–A pairing, this strategy is highly compatible with standard DNA nanostructures.

In this chapter, first I designed and synthesized **U^{OH}**-containing DNA strands for metal-responsive SDRs (Sect. 2.2). Then Gd^{III}-responsive base pair switching of **U^{OH}** was confirmed by analyzing thermal stability of duplexes with **U^{OH}**–**U^{OH}** or **U^{OH}**–A pairs and by studying hybridization preference of **U^{OH}**-containing strands (Sect. 2.3). Finally, the base pair switching of **U^{OH}** was applied to Gd^{III}-responsive control of reversible SDRs under isothermal conditions (Sect. 2.4).

2.2 Design and Synthesis of DNA Strands Containing 5-Hydroxyuracil Bases for Metal-Responsive DNA Strand Displacement Reactions

As mentioned above, traditional strand displacement reactions (SDRs) proceed efficiently with the aid of terminal toehold regions (Fig. 2.3a). Inspired by this mechanism, multiple **U^{OH}** bases were introduced at the termini of DNA strands as a Gd^{III}-binding site. The design of metal-responsive SDRs developed in this study is shown on Fig. 2.3b. In this system, **U^OH^**-containing strand **1** and its complementary strands **2** and **3** possessing **U^OH^** and A, respectively, are used. In the absence of metal ions, duplex **1·3** with **U^OH^**–A pairs would be preferentially formed. By the addition of Gd^{III} ions, the **U^OH^**–A pairs are destabilized and interstrand **U^OH^**–Gd^{III}–**U^OH^** complexes are formed. As a consequence, SDRs are induced to yield duplex **1·2** with Gd^{III}-mediated **U^OH^**–Gd^{III}–**U^OH** base pairs. Furthermore, this SDR was expected to proceed in a reverse direction by removing Gd^{III} ions with chelating agents such as EDTA, which dissociate duplex **1·2** and regenerate duplex **1·3**. A previous study found that it was necessary to incorporate at least three **U^OH^** bases in succession to form stable **U^OH^**–Gd^{III}–**U^OH^** pairs within the DNA duplex [25]. This may be because significant stabilization by **U^OH^**–Gd^{III}–**U^OH^** complexation is necessary to compensate for the structural distortion caused by metal complexation. It was also revealed that the degree of the Gd^{III}-dependent stabilization of the duplexes with **U^OH^**–**U^OH^** pairs can be enhanced by increasing the number of the **U^OH^**–**U^OH^** pairs from three to four. Therefore, in this study, four **U^OH^** bases were consecutively introduced at the termini of strands **1** and **2** (Design 1).

To enhance the efficiency of the SDR, I have designed another set of DNA strands **4**, **5**, and **6** containing an additional G or C base at the termini (Design 2) (Fig. 2.4).

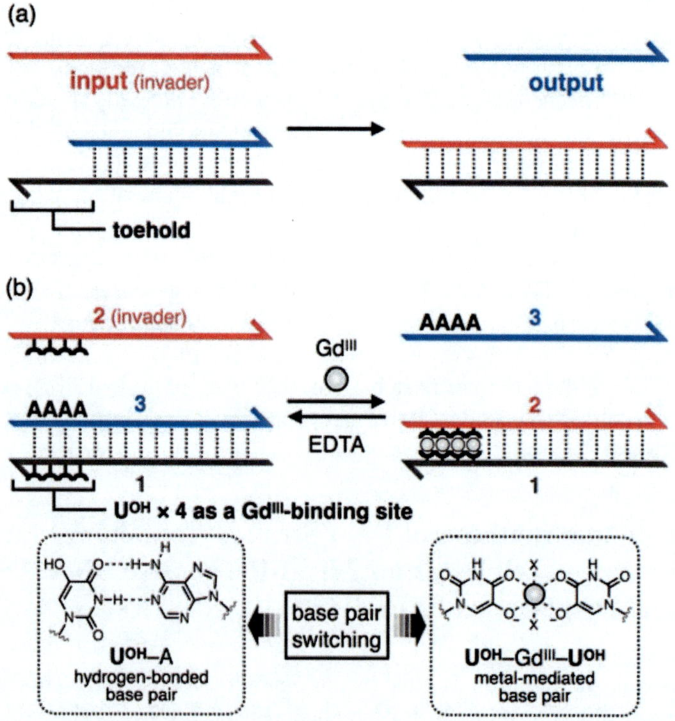

Fig. 2.3 **a** Scheme of toehold-mediated DNA strand displacement reactions (SDRs). **b** Scheme of Gd^{III}-responsive SDRs using DNA strands with four **U^OH^** bases at the termini

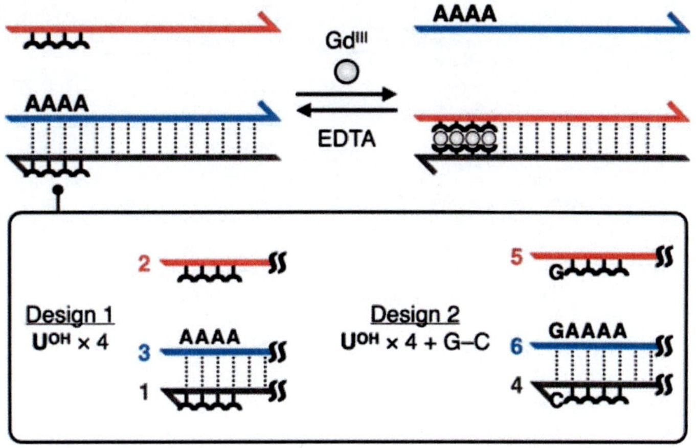

Fig. 2.4 Designs of terminal Gd^{III}-binding sequences containing **U^{OH}** bases for strand displacement reactions (SDRs). In Design 1, strands **1** and **2** contain four consecutive **U^{OH}** bases directly from the termini. In Design 2, strands **5** and **6** contain four consecutive **U^{OH}** and an additional G or C base at the termini

The introduction of a G–C pair was expected to enhance the thermal stability of duplex **4·6** with **U^{OH}**–A pairs. On the other hand, the other duplex with **U^{OH}–U^{OH}** pairs (**4·5**) would not be affected because it is not likely that the terminal G–C pair can be formed due to the neighboring **U^{OH}–U^{OH}** mismatches. Thus, it was expected that the difference in thermal stability of the two duplexes, **4·5** and **4·6**, would be increased by introduction of the terminal G–C pair. This effect would result in predominant formation of duplex **4·6** with **U^{OH}**–A pairs under Gd^{III}-free conditions.

U^{OH}-containing DNA strands were synthesized based on standard solid-phase synthesis (Fig. 2.5). **U^{OH}** phosphoramidite was prepared from 2′-deoxyuridine via a four-step reaction and was used for DNA synthesis by an automated synthesizer in the same manner as previous works [27, 28]. The sequences of the synthesized strands are summarized in Table 2.1. The length of the DNA strands was adjusted to be 16 or 17 mer, which is suitable for studying Gd^{III}-dependent changes in thermal stability of **U^{OH}**-containing DNA duplexes. Since strand **1′** does not have a natural nucleoside at the 3′ terminus, it was synthesized on Universal Support (Glen Research), with which **U^{OH}** bases were directly introduced from the 3′ end. After cleavage from the solid support and deprotection under basic conditions, the **U^{OH}**-containing strands were purified by reverse-phase HPLC. All the strands were characterized by ESI-TOF mass spectrometry (see Sect. 2.6).

Longer DNA strands (25 or 26 mer) $\mathbf{1_F}$, $\mathbf{2_D}$, $\mathbf{4_F}$, and $\mathbf{5_D}$ were designed for SDRs (Table 2.2) and synthesized by enzymatic ligation reactions using the shorter **U^{OH}**-containing strands **1′**, **2′**, **4′**, and **5′** (16 or 17 mer). To evaluate the progress of the SDRs by fluorescence measurement, a FAM fluorophore or a Dabcyl quencher was introduced into the termini of the strands. For the synthesis of strands $\mathbf{1_F}$ and $\mathbf{4_F}$, strands **1′** and **4′** were phosphorylated by T4 polynucleotide kinase (T4 PNK). The

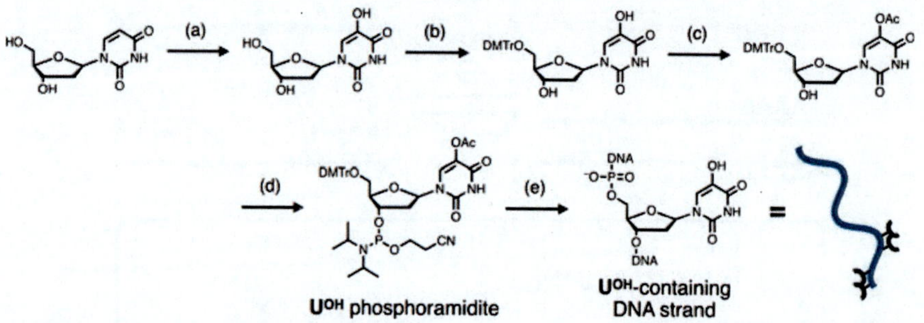

Fig. 2.5 Synthetic route for $\mathbf{U^{OH}}$ phosphoramidite and $\mathbf{U^{OH}}$-containing DNA strands. (a) (i) Br_2/H_2O, rt; (ii) pyridine, rt, 61%. (b) DMTrCl, DMAP, pyridine, rt, 74%. (c) Ac_2O, pyridine, rt, 74%. (d) $NC(CH_2)_2OPCl[N(^iPr)_2]$, DIPEA, CH_2Cl_2, rt, 38%. (e) Solid-phase DNA synthesis by an automated synthesizer. Synthesized DNA strands were cleaved from the solid support and deprotected by 25% NH_3 aq at room temperature. DMTr = 4,4′-dimethoxytrityl, DMAP = 4-dimethylaminopyridine, Ac = acetyl, ^{i}Pr = isopropyl, DIPEA = diisopropylethylamine

Table 2.1 Sequences of DNA strands synthesized by an automated synthesizer

Name	Sequence (5′ to 3′)
1′	CAC ATT GTT GTA $\mathbf{U^{OH}U^{OH}U^{OH}}$ $\mathbf{U^{OH}}$
2′	$\mathbf{U^{OH}U^{OH}U^{OH}}$ $\mathbf{U^{OH}}$TA CAA CAA TGT G
4′	CAC ATT GTT GTA $\mathbf{U^{OH}U^{OH}U^{OH}}$ $\mathbf{U^{OH}}$C
5′	G$\mathbf{U^{OH}U^{OH}}$ $\mathbf{U^{OH}U^{OH}}$T ACA ACA ATG TG

phosphorylated donor strands were ligated with FAM-modified primer by using T4 DNA ligase in the presence of template strands. Purification by denaturing polyacrylamide gel electrophoresis (PAGE) afforded the target strands. Dabcyl-labeled strands $\mathbf{2_D}$ and $\mathbf{5_D}$ were synthesized by the enzymatic ligation of strands **2′** and **5′** with a phosphorylated natural strand. The synthesized $\mathbf{U^{OH}}$-containing strands were characterized by MALDI-TOF mass spectrometry (see Sect. 2.6).

In summary, DNA strands for Gd^{III}-triggered SDRs were designed by introducing four $\mathbf{U^{OH}}$ bases at the termini. Based on the designed sequences, short $\mathbf{U^{OH}}$-containing DNA strands (16 or 17 mer) were synthesized by solid-phase DNA synthesis using an automated synthesizer. Furthermore, longer strands for the SDRs were synthesized by enzymatic ligation reactions. Since all desired strands were

Table 2.2 Sequences of DNA strands synthesized by enzymatic ligation

Name	Sequence (5′ to 3′)
1_F	FAM—AGC TAC CTC CAC ATT GTT GTA $\mathbf{U^{OH}U^{OH}U^{OH}}$ $\mathbf{U^{OH}}$
2_D	$\mathbf{U^{OH}U^{OH}U^{OH}}$ $\mathbf{U^{OH}}$TA CAA CAA TGT GGA GGT AGC T—Dabcyl
4_F	FAM—AGC TAC CTC CAC ATT GTT GTA $\mathbf{U^{OH}U^{OH}U^{OH}}$ $\mathbf{U^{OH}}$C
5_D	G$\mathbf{U^{OH}U^{OH}}$ $\mathbf{U^{OH}U^{OH}}$T ACA ACA ATG TGG AGG TAG CT—Dabcyl

successfully obtained with high purity, they were expected to be applicable for studying the base pair switching of $\mathbf{U^{OH}}$ and Gd^{III}-triggered SDRs.

2.3 Gd^{III}-Responsive Base Pair Switching of 5-Hydroxyuracil Bases

As discussed in Sect. 2.2, the strand displacement reaction (SDR) was expected to be triggered by Gd^{III}-responsive base pair switching from $\mathbf{U^{OH}}$–A to $\mathbf{U^{OH}}$–Gd^{III}–$\mathbf{U^{OH}}$. Herein the base pair switching of $\mathbf{U^{OH}}$ was evaluated by comparing thermal stability of a duplex containing $\mathbf{U^{OH}}$–A pairs and one containing $\mathbf{U^{OH}}$–$\mathbf{U^{OH}}$ pairs. The Gd^{III}-dependent base pair switching was further demonstrated by controlling the hybridization preference of $\mathbf{U^{OH}}$-containing strands. In addition to the $\mathbf{U^{OH}}$-containing strands synthesized in Sect. 2.2, several natural strands containing A or T bases instead of $\mathbf{U^{OH}}$ were also used (Table 2.3).

2.3.1 Gd^{III}-Dependent Hybridization Preference of U^{OH}-containing DNA Strands

To study hybridization preference of $\mathbf{U^{OH}}$-containing strands in the absence and presence of Gd^{III} ions, thermal stability of DNA duplexes with $\mathbf{U^{OH}}$–$\mathbf{U^{OH}}$ or $\mathbf{U^{OH}}$–A pairs was estimated by duplex melting analysis. The result showed that duplex **1′·2′** with four $\mathbf{U^{OH}}$–$\mathbf{U^{OH}}$ pairs at the terminus ($T_m = 37.9$ °C) was significantly stabilized by the addition of 4 equiv. of Gd^{III} ions ($T_m = 64.1$ °C, $\Delta T_m = +26.1$ °C) (Fig. 2.6a). In contrast, duplex **1T′·2T′** containing T–T pairs instead of $\mathbf{U^{OH}}$–$\mathbf{U^{OH}}$ pairs did not show any changes in the thermal stability (Fig. 2.6b). This Gd^{III}-dependent duplex stabilization was compatible with the previous study [25], in which the formation of three consecutive $\mathbf{U^{OH}}$–Gd^{III}–$\mathbf{U^{OH}}$ pairs in the middle of the DNA duplex was

Table 2.3 Sequences of DNA strands used in Sect. 2.3

Name	Sequence (5′ to 3′)
1′	CAC ATT GTT GTA $\mathbf{U^{OH}U^{OH}U^{OH}}$ $\mathbf{U^{OH}}$
1T′	CAC ATT GTT GTA **TTT T**
2′	$\mathbf{U^{OH}U^{OH}U^{OH}}$ $\mathbf{U^{OH}}$TA CAA CAA TGT G
2T′	**TTT T**TA CAA CAA TGT G
3′	**AAA A**TA CAA CAA TGT G
4′	CAC ATT GTT GTA $\mathbf{U^{OH}U^{OH}U^{OH}}$ $\mathbf{U^{OH}}$C
5′	G$\mathbf{U^{OH}U^{OH}}$ $\mathbf{U^{OH}U^{OH}}$T ACA ACA ATG TG
6′	G**AA AA**T ACA ACA ATG TG

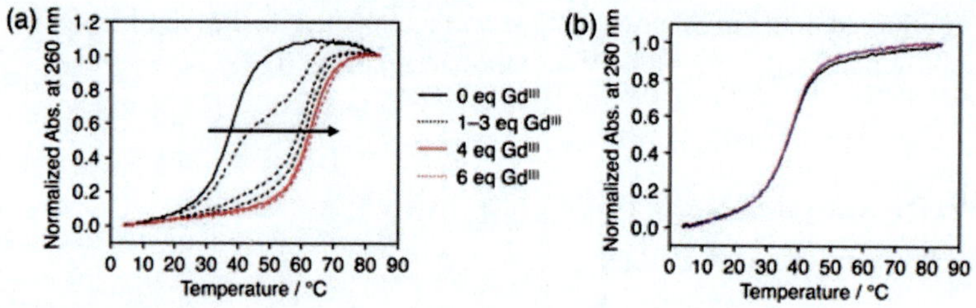

Fig. 2.6 Melting curves of DNA duplexes containing **a** **U^{OH}–U^{OH}** (duplex **1′·2′**) and **b** T–T pairs (duplex **1T′·2T′**) in the absence and presence of Gd^{III} ions. [duplex] = 2.0 μM, [Gd^{III}]/[duplex] = 0 (black solid lines), 1, 2, 3 (black dotted lines), 4 (red or purple solid line), and 6 (red dotted line) in 10 mM HEPES–NaOH buffer (pH 8.0), 100 mM NaCl, 0.2 °C/min

confirmed. Thus, it was concluded that this significant stabilization of duplex **1′·2′** was caused by the interstrand **U^{OH}**–Gd^{III}–**U^{OH}** complexation.

The metal complexation was further confirmed by UV–vis absorption spectroscopy (Fig. 2.7a). The absorption spectrum of duplex **1′·2′** showed that the addition of Gd^{III} ions resulted in gradual increase in the absorbance at 310 nm. This absorbance change can be attributed to the deprotonation of the 5-OH group of the **U^{OH}** bases upon Gd^{III} binding [25]. The metal complexation was also studied by CD spectroscopy (Fig. 2.7b). In the absence of Gd^{III} ions, the CD spectrum showed positive and negative Cotton effects around 280 and 250 nm, respectively, characteristic of the normal DNA duplex with a right-handed B-type helix. When Gd^{III} ions were added, gradual decrease in the CD intensity was observed around 280 nm. This result suggests structural distortion of the DNA duplexes due to the **U^{OH}**–Gd^{III}–**U^{OH}** complexation. The **U^{OH}**–Gd^{III}–**U^{OH}** complexation was also indicated by the increase in the CD intensity around 325 nm. The changes in UV absorption and CD intensity were converged at 4 equiv. of Gd^{III} ions. Thus, it was suggested that four Gd^{III} ions were incorporated into duplex **1′·2′** by the quantitative formation of **U^{OH}**–Gd^{III}–**U^{OH}** pairs. The duplex with four Gd^{III} ions was also observed by ESI mass spectrometry (Fig. 2.7c). Since Gd^{III} complexes generally prefer eight or more coordination numbers, it is highly possible that additional ligands such as H_2O and neighboring nucleobases coordinate to the Gd^{III} center.

In contrast, duplex **1′·3′** with four **U^{OH}**–A pairs (T_m = 45.3 °C) was thermally destabilized in the presence of Gd^{III} ions (T_m = 38.9 °C, ΔT_m = –6.4 °C) (Fig. 2.8a). However, the melting profile of duplex **1T′·3′** with four T–A pairs instead of **U^{OH}**–A was not affected by the addition of Gd^{III} ions (Fig. 2.8b). These results suggested that duplex **1′·3′** was thermally destabilized possibly because the hydrogen bonding between **U^{OH}** and A bases was weakened by the binding of Gd^{III} to **U^{OH}**.

Melting temperatures of two duplexes **1′·2′** and **1′·3′** are summarized in Table 2.4. In the absence of Gd^{III}, duplex **1′·3′** with **U^{OH}**–A pairs was more stable than duplex **1′·2′** containing **U^{OH}**–**U^{OH}** mismatches. However, duplex **1′·2′** became much more stable in the presence of 4 equiv. of Gd^{III} because of the interstrand **U^{OH}**–Gd^{III}–**U^{OH}** complexation. Accordingly, the relative stability of the two duplexes was inverted

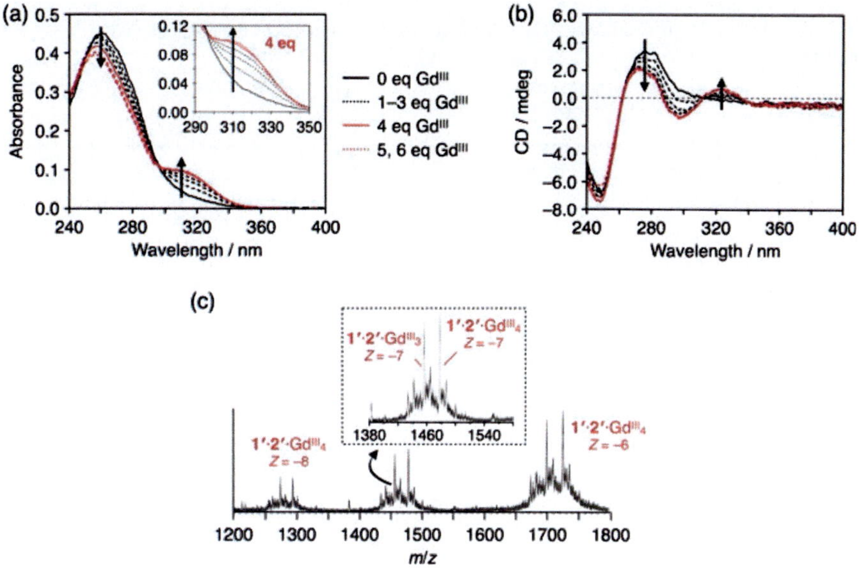

Fig. 2.7 $\mathbf{U^{OH}}$–Gd^{III}–$\mathbf{U^{OH}}$ complexation within duplex **1′·2′** containing four $\mathbf{U^{OH}}$–$\mathbf{U^{OH}}$ pairs. **a** UV–vis absorption spectra and **b** circular dichroism (CD) spectra of duplex **1′·2′** in the presence of varying amounts of Gd^{III} ions. [duplex] = 2.0 μM, [Gd^{III}]/[duplex] = 0 (black solid lines), 1, 2, 3 (black dotted lines), 4 (red solid lines), 5, and 6 (red dotted lines) in 10 mM HEPES–NaOH buffer (pH 8.0), 100 mM NaCl. l = 1.0 cm, 4 °C. **c** ESI mass spectrum of duplex **1′·2′** in the presence of Gd^{III} ions (4 equiv.). [duplex] = 40 μM, [Gd^{III}]/[duplex] = 4 in 20 mM NH_4OAc buffer (pH 7.0), negative mode. Signals with three Gd^{III} ions were also observed possibly because the terminal Gd^{III} ion was dissociated during the measurement

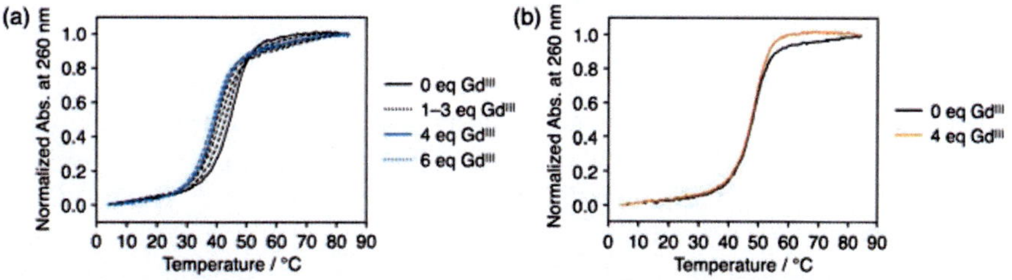

Fig. 2.8 Melting curves of DNA duplexes containing **a** $\mathbf{U^{OH}}$–A (duplex **1′·3′**) and **b** T–A (duplex **1 T′·3′**) pairs in the absence and presence of Gd^{III} ions. [duplex] = 2.0 μM, [Gd^{III}]/[duplex] = 0 (black solid lines), 1, 2, 3 (black dotted lines), 4 (blue or orange solid line), and 6 (blue dotted line) in 10 mM HEPES–NaOH buffer (pH 8.0), 100 mM NaCl, 0.2 °C/min

by the addition of Gd^{III} ions. This result indicates that base pair switching between $\mathbf{U^{OH}}$–A and $\mathbf{U^{OH}}$–Gd^{III}–$\mathbf{U^{OH}}$ pairs can occur in a Gd^{III}-dependent manner.

Next, the hybridization preference of $\mathbf{U^{OH}}$-containing strand **1′** was evaluated. An equimolar mixture of strands **1′**, **2′**, and **3′** was annealed in the absence and presence of Gd^{III} ions, and the hybridization products were analyzed by native polyacrylamide gel

Table 2.4 Melting temperatures (T_m) of $\mathbf{U^{OH}}$-containing duplexes in the absence and presence of Gd^{III} ions[a]

Gd^{III}/eq	Duplex **1′·2′** ($\mathbf{U^{OH}}$–$\mathbf{U^{OH}}$)		Duplex **1′·3′** ($\mathbf{U^{OH}}$–A)	
	T_m/°C	ΔT_m/°C	T_m/°C	ΔT_m/°C
0	37.9 ± 0.4 °C	–	45.3 ± 0.5 °C	–
4	64.1 ± 0.7 °C	+ 26.1 ± 0.9 °C	38.9 ± 0.1 °C	– 6.4 ± 0.5 °C

[a]Averaged values calculated from three independent measurements are shown.
Error values stand for standard errors.

electrophoresis (PAGE) (Fig. 2.9a). Strand **3′** was labeled with a FAM fluorophore to quantitatively evaluate which duplex, **1′·2′** or **1′·3′**, was formed. As shown in Fig. 2.9b, duplex **1′·3′** was predominantly formed in the absence of Gd^{III} ions (81%). However, in the presence of 4 equiv. of Gd^{III} ions, dissociated single strand **3′** was mainly observed (90%), indicating the formation of duplex **1′·2′** containing the $\mathbf{U^{OH}}$–Gd^{III}–$\mathbf{U^{OH}}$ pairs. In fact, Gd^{III}-containing duplex **1′·2′** was detected after staining the gel by SYBR Gold (Fig. 2.9c). Figure 2.9d shows the result of Gd^{III} titration experiments. The yield of duplex **1′·2′** was gradually enhanced by increasing the amount of Gd^{III} ions from 0 to 4 equiv. and almost unchanged in the presence of excess Gd^{III} ions. These results clearly showed that the addition of Gd^{III} ions altered the hybridization preference of $\mathbf{U^{OH}}$-containing strand **1′** by Gd^{III}-mediated base pair switching from $\mathbf{U^{OH}}$–A to $\mathbf{U^{OH}}$–Gd^{III}–$\mathbf{U^{OH}}$.

2.3.2 Gd^{III}-Responsive Base Pair Switching under Isothermal Conditions

Further investigation proved that the hybridization partner of $\mathbf{U^{OH}}$-containing strand **1′** can be switched in a Gd^{III}-dependent manner even under isothermal conditions (Fig. 2.10a). Gd^{III} ions (4 equiv.) were added to the preannealed mixture of strands **1′**, **2′**, and **3′** to induce the strand exchange from duplex **1′·3′** to **1′·2′**. Native PAGE analysis revealed that duplex **1′·2′** was formed in 79% yield after 4-h incubation at 25 °C, which was almost the same level as the sample annealed in the presence of Gd^{III} ions (90%) (Fig. 2.10b). As shown in time-course analysis in Fig. 2.10c, the strand exchange was almost completed within 1 h.

The exchange of the hybridization partner was found to be reversible by removing Gd^{III} ions with chelating agents, EDTA (Fig. 2.11a). To a preannealed mixture of strands **1′**, **2′**, and **3′** with 4 equiv. of Gd^{III} ions, 4 equiv. of EDTA were added. As a result, duplex **1′·3′** was regenerated in 66% yield after 8 h (Fig. 2.11b). The time-course analysis shown in Fig. 2.11c revealed that the strand exchange was almost completed within 8 h.

Finally, iterative switching of the hybridization partner was examined by the repeated addition of Gd^{III} ions and EDTA under isothermal conditions (Fig. 2.12a).

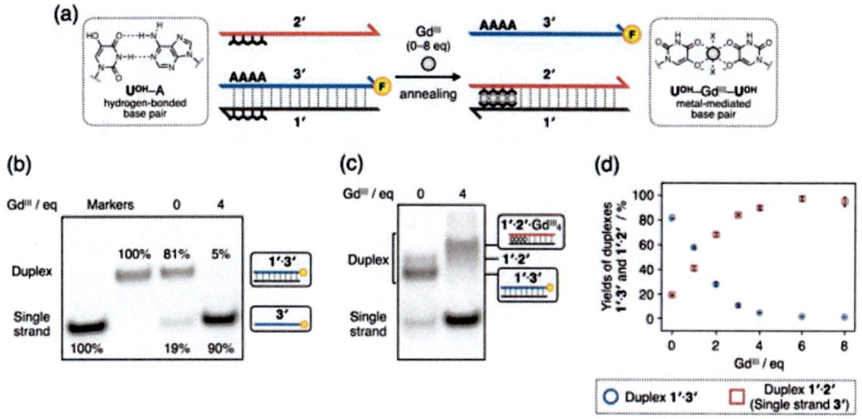

Fig. 2.9 Gd^{III}-dependent control of the hybridization preference of **U^{OH}**-containing strand **1′** (16 mer). **a** Scheme. "X" denotes additional ligands coordinating to the Gd^{III} ion such as H_2O and neighboring nucleobases. F: FAM. **b**, **c** Native PAGE analysis of an equimolar mixture of strands **1′**, **2′**, and **3′** (16 mer each) in the absence and presence of Gd^{III} ions. **b** FAM detection, **c** after SYBR Gold staining. **d** The yields of duplex **1′·3′** and **1′·2′** in the presence of different concentrations of Gd^{III} ions. [each DNA] = 2.0 μM, [Gd^{III}]/[each DNA] = 0, 1, 2, 3, 4, 6, or 8 in 10 mM HEPES–NaOH buffer (pH 8.0), 100 mM NaCl. The samples were annealed prior to the analysis. 20% gel at 4 °C. The yields were calculated from the average of at least three independent experiments by comparing the band intensities on the gel with those of the markers

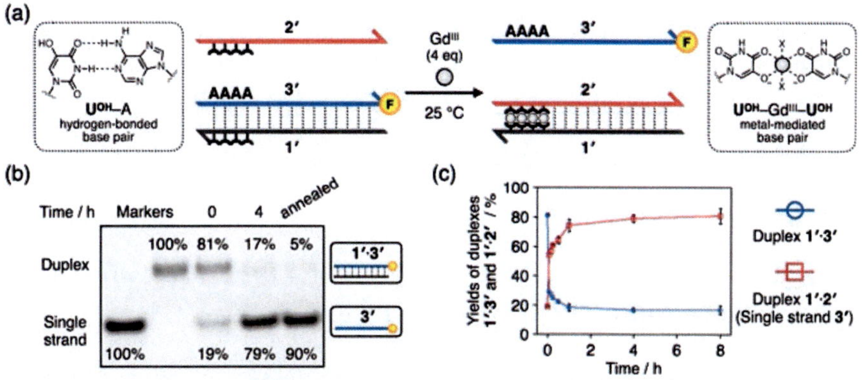

Fig. 2.10 Gd^{III}-triggered exchange of the hybridization partner of **U^{OH}**-containing strand **1′** (16 mer) under isothermal conditions. **a** Scheme. "X" denotes additional ligands coordinating to the Gd^{III} ion such as H_2O and neighboring nucleobases. F: FAM. **b** Native PAGE analysis of an equimolar mixture of strands **1′**, **2′**, and **3′** (16 mer each) before and after the addition of Gd^{III} ions. **c** Time-course changes in the yields of duplexes **1′·3′** and **1′·2′** after the addition of Gd^{III} ions. [each DNA] = 2.0 μM, [Gd^{III}]/[each DNA] = 4 in 10 mM HEPES–NaOH buffer (pH 8.0), 100 mM NaCl. After annealing an equimolar mixture of strands **1′**, **2′**, and **3′**, 4 equiv. of Gd^{III} ions were added and the sample was incubated at 25 °C. 20% gel at 4 °C. The yields were calculated from the average of three independent experiments by comparing the band intensities on the gel with those of the markers. Error bars stand for standard errors

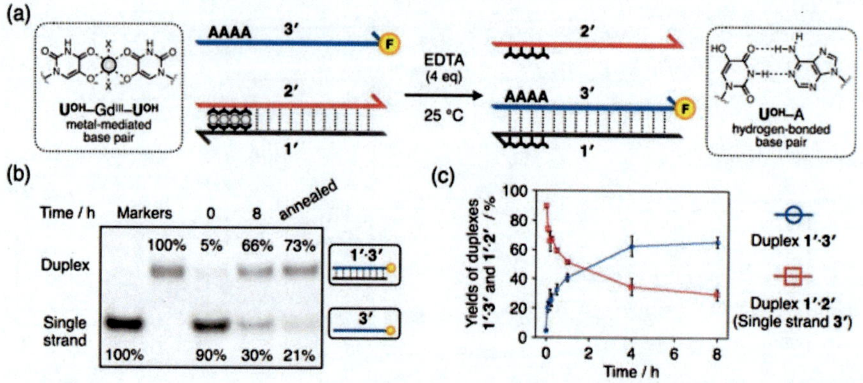

Fig. 2.11 EDTA-triggered regeneration of duplex **1′·3′** under isothermal conditions. **a** Scheme. "X" denotes additional ligands coordinating to the Gd^{III} ion such as H_2O and neighboring nucleobases. F: FAM. **b** Native PAGE analysis of a mixture of strands **1′**, **2′**, and **3′** (16 mer, 1 equiv. each) and Gd^{III} (4 equiv.) before and after the addition of EDTA. **c** Time-course changes in the yields of duplexes **1′·3′** and **1′·2′** after the addition of EDTA. [each DNA] = 2.0 μM, [Gd^{III}]/[each DNA] = 4, [EDTA]/[each DNA] = 4 in 10 mM HEPES–NaOH buffer (pH 8.0), 100 mM NaCl. After annealing a mixture of strands **1′**, **2′**, and **3′** (1 equiv. each) and Gd^{III} (4 equiv.), 4 equiv. of EDTA were added and the sample was incubated at 25 °C. 20% gel at 4 °C. The yields were calculated from the average of three independent experiments by comparing the band intensities on the gel with those of the markers. Error bars stand for standard errors

Gd^{III} ions and EDTA (4 equiv. each) were alternately added in 2-h intervals at a constant temperature of 25 °C. Native PAGE analysis showed that the hybridization partner of strand **1′** was reversibly exchanged (Fig. 2.12b, c). Accordingly, it was confirmed that the hybridization preference of **U^{OH}**-containing strand **1′** can be reversibly controlled based on the Gd^{III}-responsive base pair switching of **U^{OH}**.

2.3.3 Effects of the Introduction of an Additional G–C Pair at the Termini of DNA Duplexes

The hybridization preference of strand **1′** was successfully controlled in a Gd^{III}-dependent manner as described in Sects. 2.3.1 and 2.3.2. However, the formation of duplexes **1′·3′** and **1′·2′** in the absence and presence of Gd^{III} ions, respectively, was not quantitative. In particular, duplex **1′·3′** was formed in only 81% yield under Gd^{III}-free conditions. To enhance the formation of duplexes containing **U^{OH}**–A pairs in the absence of Gd^{III} ions, DNA strands **4′**, **5′**, and **6′** with an additional G–C pair at the termini (Design 2) were utilized. Thermal stability of duplexes **4′·5′** with **U^{OH}–U^{OH}** pairs and **4′·6′** with **U^{OH}**–A pairs were evaluated in the same manner as Design 1. Melting experiments showed that duplexes **4′·5′** and **4′·6′** exhibited Gd^{III}-dependent thermal stabilization and destabilization, respectively, in a manner similar to duplexes **1′·2′** and **1′·3′** in Design 1 (Fig. 2.13). The calculated melting temperatures are shown

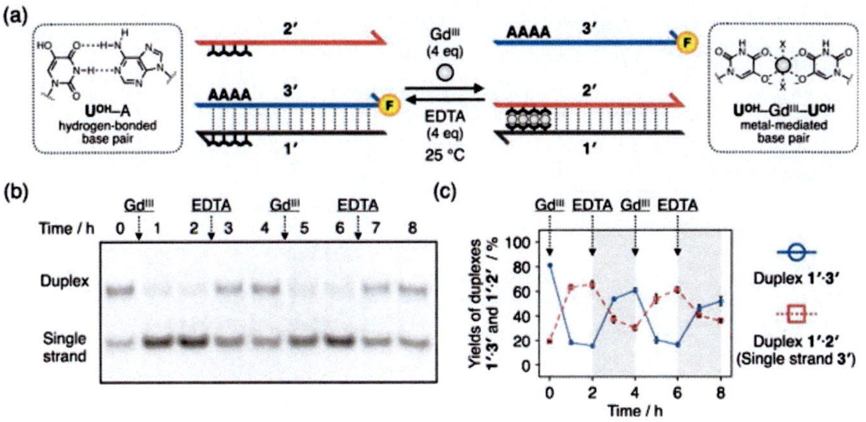

Fig. 2.12 Reversible exchange of the hybridization partner of $\mathbf{U^{OH}}$-containing strand **1′** (16 mer) by the sequential addition of Gd^{III} ions and EDTA under isothermal conditions. **a** Scheme. "X" denotes additional ligands coordinating to the Gd^{III} ion such as H_2O and neighboring nucleobases. F: FAM. **b** Native PAGE analysis of the reversible strand exchange in response to Gd^{III} ions and EDTA. **c** Time-course changes in the yields of duplexes **1′·3′** and **1′·2′** during the repeated strand exchange. [each DNA] = 2.0 μM, [Gd^{III}]/[each DNA] = 4, [EDTA]/[each DNA] = 4 in 10 mM HEPES–NaOH buffer (pH 8.0), 100 mM NaCl. After annealing an equimolar mixture of strands **1′**, **2′**, and **3′**, 4 equiv. of Gd^{III} ions and EDTA were repeatedly added and the sample was incubated at 25 °C. 20% gel at 4 °C. The yields were calculated from the average of three independent experiments by comparing the band intensities on the gel with those of the markers. Error bars stand for standard errors

in Table 2.5. The melting temperatures of duplex **4′·5′** were 36.7 °C and 64.0 °C in the absence and presence of Gd^{III} ions (4 equiv.), respectively, which were almost comparable with those of duplex **1′·2′** without a G–C pair. On the other hand, duplex **4′·6′** with $\mathbf{U^{OH}}$–A pairs showed slightly higher stability (T_m = 46.7 °C) than duplex **1′·3′** (45.3 °C). As a result, the degree of the Gd^{III}-dependent thermal destabilization of duplex **4′·6′** (ΔT_m = − 8.1 °C) was enhanced compared with that of duplex **1′·3′** (− 6.4 °C).

Melting temperatures of the duplexes in Design 1 and 2 are compared in Fig. 2.14. Under Gd^{III}-free conditions (Fig. 2.14b), duplex **1′·3′** with $\mathbf{U^{OH}}$–A pairs was 7.4 °C more stable than duplex **1′·2′** with $\mathbf{U^{OH}}$–$\mathbf{U^{OH}}$ pairs (Design 1). On the other hand, duplex **4′·6′** was 10.0 °C more stable than duplex **4′·5′** (Design 2), indicating that the difference of the T_m values of the two duplexes was increased by introducing a G–C pair. Accordingly, it was suggested that the formation of the duplex with $\mathbf{U^{OH}}$–A pairs would be more preferable in Design 2. In the presence of 4 equiv. of Gd^{III} ions (Fig. 2.14c), no significant changes in the T_m differences were observed between Design 1 and 2.

The switching efficiency of the hybridization preference of the $\mathbf{U^{OH}}$-containing strands was improved in Design 2 compared with that of Design 1 (Fig. 2.15). In Design 2, duplex **4′·6′** with $\mathbf{U^{OH}}$–A pairs was formed in 89% yield, which was higher than that of Design 1 (81%) (Fig. 2.15b). In the presence of 4 equiv. of Gd^{III} ions,

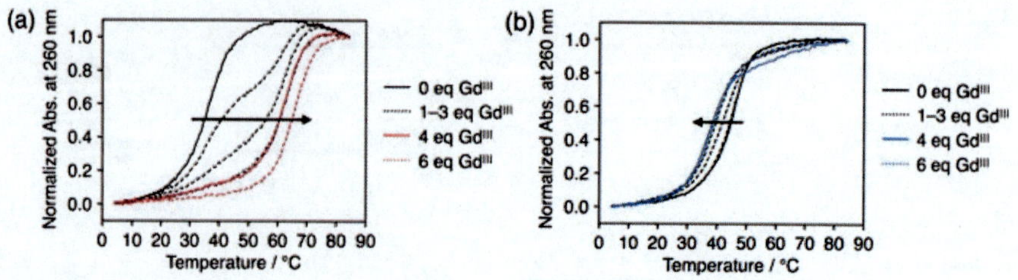

Fig. 2.13 Melting curves of DNA duplexes containing **a** U^{OH}–U^{OH} pairs (duplex **4′·5′**) and **b** U^{OH}–A pairs (duplex **4′·6′**) in the absence and presence of Gd^{III} ions. An additional G–C pair was introduced into the termini. [duplex] = 2.0 μM, [Gd^{III}]/[duplex] = 0 (black solid lines), 1, 2, 3 (black dotted lines), 4 (red or blue solid line), and 6 (red or blue dotted line) in 10 mM HEPES–NaOH buffer (pH 8.0), 100 mM NaCl, 0.2 °C/min

Table 2.5 Melting temperatures (T_m) of U^{OH}-containing duplexes with additional G–C pairs (Design 2) in the absence and presence of Gd^{III} ions[a]

Gd^{III}/eq	Duplex **4′·5′** (U^{OH}–U^{OH})		Duplex **4′·6′** (U^{OH}–A)	
	T_m/°C	ΔT_m/°C	T_m/°C	ΔT_m/°C
0	36.7 ± 1.4 °C	–	46.7 ± 1.0 °C	–
4	64.0 ± 0.7 °C	+ 27.4 ± 0.7 °C	38.6 ± 0.4 °C	– 8.1 ± 1.4 °C

[a] Averaged values calculated from three independent measurements are shown.
Error values stand for standard errors.

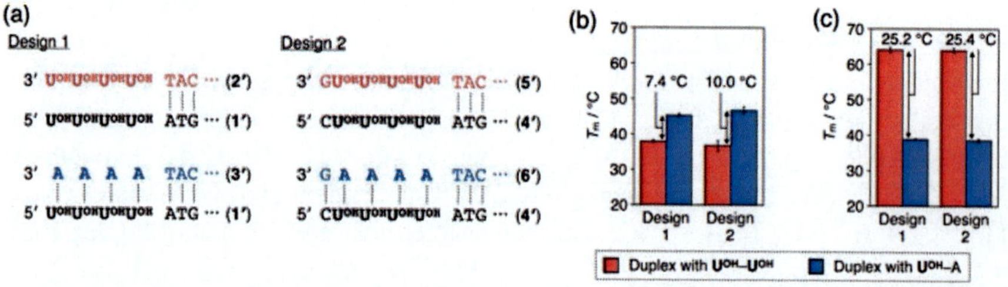

Fig. 2.14 Comparison of the melting temperatures of DNA duplexes between Design 1 (without G–C) and 2 (with G–C). **a** Sequence designs. **b**, **c** Melting temperatures of duplexes containing U^{OH}–U^{OH} or U^{OH}–A pairs. **b** Without Gd^{III} ions, **c** with 4 equiv. of Gd^{III} ions. Error bars stand for standard errors

the yields of the duplex with U^{OH}–Gd^{III}–U^{OH} pairs were 90% for both sequence designs (Fig. 2.15c). Therefore, the preferential formation of the duplex with U^{OH}–Gd^{III}–U^{OH} was retained even after the introduction of a G–C pair. Overall, it can be concluded that the switching efficiency of the hybridization preferences was slightly improved by introducing an additional G–C pair into the termini of DNA duplexes.

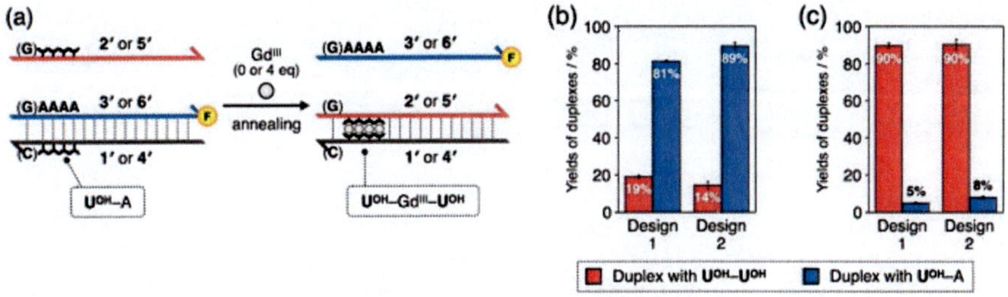

Fig. 2.15 Comparison of the yields of DNA duplexes with $\mathbf{U^{OH}}$–$\mathbf{U^{OH}}$ and $\mathbf{U^{OH}}$–A pairs in the Gd^{III}-dependent switching of the hybridization partner of the $\mathbf{U^{OH}}$-containing strands (**1′** or **4′**). Results of Design 1 (without G–C) and Design 2 (with G–C) are shown. **a** Sequence design. Strands **4′**, **5′**, and **6′** contain an additional G or C base at the termini. F: FAM. **b**, **c** Yields of the duplexes. **b** Without Gd^{III} ions, **c** with 4 equiv. of Gd^{III} ions. Error bars stand for standard errors

2.4 Metal-Responsive DNA Strand Displacement Reactions Driven by Base Pair Switching of 5-Hydroxyuracil Bases

Gd^{III}-responsive base pair switching of the $\mathbf{U^{OH}}$ base was demonstrated and applied to control the hybridization preference of $\mathbf{U^{OH}}$-containing strands, as described in Sect. 2.3. Based on these results, it was expected that the Gd^{III}-dependent base pair switching would be further applicable to DNA strand displacement reactions (SDRs). I have investigated Gd^{III}-triggered SDRs using longer $\mathbf{U^{OH}}$-containing strands **1** and **2** (25 mer) (Fig. 2.16a). Since starting duplex **1**·**3** with $\mathbf{U^{OH}}$–A pairs is long enough, the SDR was expected to proceed via binding of invader strand **2** in a manner similar to the conventional toehold-mediated SDRs using natural strands (Fig. 2.16b). Subsequent branch migration results in the release of strand **3**. Sequences of DNA strands used in this section are shown in Table 2.6. Strands $\mathbf{2_D}$ and **3** containing $\mathbf{U^{OH}}$ and A bases, respectively, were designed to be complementary to $\mathbf{U^{OH}}$-containing strand $\mathbf{1_F}$. Since the introduction of an additional G–C pair improved the efficiency of the strand exchange as shown in Sect. 2.3, strands $\mathbf{4_F}$, $\mathbf{5_D}$, and **6** with the terminal C and G bases were prepared as well. The $\mathbf{U^{OH}}$-containing strands were synthesized by enzymatic ligation reactions, as described in Sect. 2.2. In addition, several strands containing T bases instead of $\mathbf{U^{OH}}$ were also used for control experiments.

The Gd^{III}-triggered SDR was first examined by the addition of Gd^{III} ions to a mixture of duplex $\mathbf{1_F}$·**3** containing the $\mathbf{U^{OH}}$–A pairs and invader strand $\mathbf{2_D}$ with the $\mathbf{U^{OH}}$ bases (Fig. 2.17a). To evaluate the progress of the SDR by fluorescence measurement, strands $\mathbf{1_F}$ and $\mathbf{2_D}$ were labeled by a FAM fluorophore and a Dabcyl quencher, respectively. The formation of duplex $\mathbf{1_F}$·$\mathbf{2_D}$ can be indicated by quenching of the emission from the FAM. Time-course fluorescence measurement revealed that the emission intensity from the FAM was gradually decreased after the addition of Gd^{III} ions (Fig. 2.17b, red solid line). This result indicates that the SDR proceeded to give duplex $\mathbf{1_F}$·$\mathbf{2_D}$ with a pair of FAM and Dabcyl at its terminus. When strands

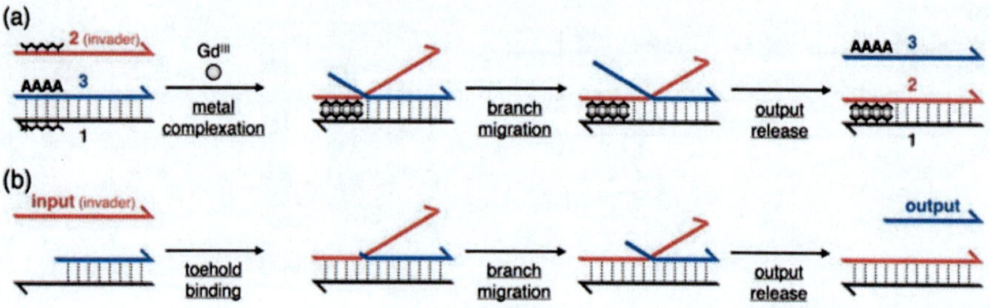

Fig. 2.16 **a** Mechanism of Gd^{III}-triggered strand displacement reactions (SDRs) driven by base pair switching of $\mathbf{U^{OH}}$. **b** Mechanism of conventional toehold-mediated SDRs using natural DNA strands

Table 2.6 Sequences of DNA strands used in Sect. 2.4

Name	Sequence (5′ to 3′)
1_F	FAM—AGC TAC CTC CAC ATT GTT GTA $\mathbf{U^{OH}U^{OH}U^{OH}\ U^{OH}}$
$1T_F$	FAM—AGC TAC CTC CAC ATT GTT GTA **TTT T**
2_D	$\mathbf{U^{OH}U^{OH}U^{OH}\ U^{OH}}$TA CAA CAA TGT GGA GGT AGC T—Dabcyl
3	**AAA A**TA CAA CAA TGT GGA GGT AGC T
4_F	FAM—AGC TAC CTC CAC ATT GTT GTA $\mathbf{U^{OH}U^{OH}U^{OH}\ U^{OH}}$C
$4T_F$	FAM—AGC TAC CTC CAC ATT GTT GTA **TTT T**C
5_D	G$\mathbf{U^{OH}U^{OH}\ U^{OH}U^{OH}}$T ACA ACA ATG TGG AGG TAG CT—Dabcyl
$5T_D$	G**TT TTT** ACA ACA ATG TGG AGG TAG CT—Dabcyl
5s	TAC AAC AAT GTG GAG GTA GCT
6	G**AA AAT** ACA ACA ATG TGG AGG TAG CT
6_D	G**AA AAT** ACA ACA ATG TGG AGG TAG CT—Dabcyl

$\mathbf{4_F}$, $\mathbf{5_D}$, and **6** containing additional G or C bases were used, the Gd^{III}-triggered SDR proceeded in a similar rate (Fig. 2.17b, red dotted line). When the $\mathbf{U^{OH}}$ bases in strands $\mathbf{4_F}$ and $\mathbf{5_D}$ were replaced by natural thymine (strands $\mathbf{4T_F}$ and $\mathbf{5T_D}$), the SDR did not proceed at all (Fig. 2.17b, black dotted line). These results clearly suggest that the SDR was driven by the Gd^{III}-responsive base pair switching from the $\mathbf{U^{OH}}$–A to $\mathbf{U^{OH}}$–Gd^{III}–$\mathbf{U^{OH}}$ pairs. Furthermore, the speed of the Gd^{III}-triggered SDR was compared with that of the conventional toehold-mediated SDRs (Fig. 2.17c). By the addition of strand $\mathbf{6_D}$ to duplex $\mathbf{4T_F{\cdot}5s}$ with a 5-nt toehold, the SDR smoothly proceeded to show immediate quenching of the emission from the FAM (Fig. 2.17b, black solid line). It was revealed that the speed of the Gd^{III}-triggered SDR was slower than that of the toehold-mediated SDR.

Selectivity of metal ions as the input for the SDRs using the $\mathbf{U^{OH}}$-containing DNA strands was also examined. To the preannealed duplex $\mathbf{1_F{\cdot}3}$, strand $\mathbf{2_D}$ and various metal ions were added to initiate the SDR. As a result, the fluorescence intensity was significantly decreased only in the presence of Gd^{III} or Eu^{III} ions, indicating high

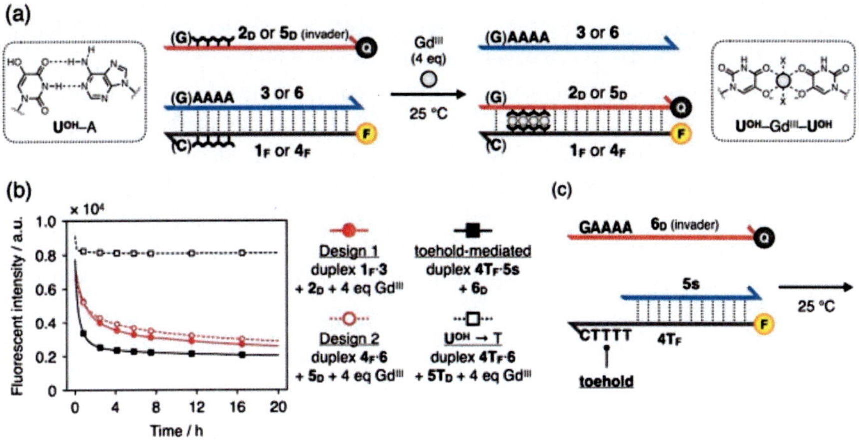

Fig. 2.17 Gd^{III}-triggered strand displacement reactions (SDRs) driven by base pair switching of the $\mathbf{U^{OH}}$ bases. **a** Scheme. "X" denotes additional ligands coordinating to the Gd^{III} ion such as H_2O and neighboring nucleobases. F: FAM, Q: Dabcyl. **b** Time-course fluorescence analysis of the SDRs. **c** Scheme of the toehold-mediated SDRs using natural DNA strands. [each DNA] = 2.0 μM, [Gd^{III}]/[each DNA] = 0 or 4 in 10 mM HEPES–NaOH buffer (pH 8.0), 100 mM NaCl. To the preannealed duplexes, 1 equiv. of the invader strands and 4 equiv. of Gd^{III} ions were added and the samples were incubated at 25 °C. For the toehold-mediated SDR, only the invader strand was added. λ_{ex} = 495 nm, λ_{em} = 519 nm. The fluorescence intensity was recorded every 10 min

selectivity of the SDR toward lanthanide ions (Fig. 2.18). This result is consistent with the previous study reporting that $\mathbf{U^{OH}}$ bases selectively form stable metal-mediated base pairs with lanthanide ions [25].

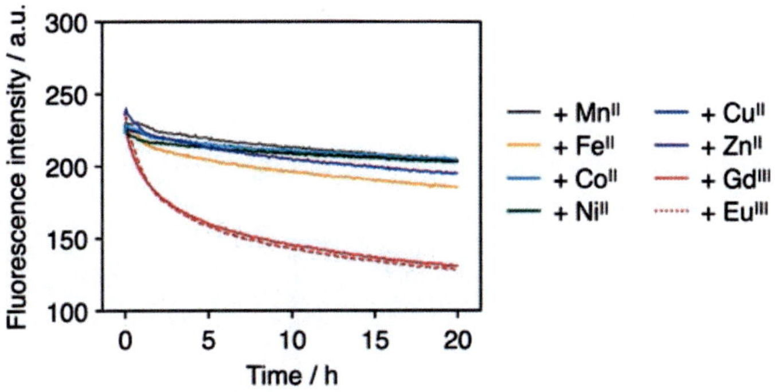

Fig. 2.18 Time-course fluorescence changes during the SDRs using $\mathbf{U^{OH}}$-containing DNA strands in the presence of various metal ions as the input. [each DNA] = 2.0 μM, [metal]/[each DNA] = 4 in 10 mM HEPES–NaOH buffer (pH 8.0), 100 mM NaCl. To the preannealed duplex ($\mathbf{1_F}$·**3**), 1 equiv. of the invader strand ($\mathbf{2_D}$) and 4 equiv. of metal ions were added. The samples were incubated at 25 °C. The fluorescence intensity was recorded every 10 min. λ_{ex} = 495 nm, λ_{em} = 520 nm

To study the detailed mechanism of the Gd^{III}-triggered SDRs, the speed of metal complexation during the SDRs was examined by UV–vis absorption measurement (Fig. 2.19). After the addition of Gd^{III} ions to 17-mer duplex **4′·5′**, the absorbance at 310 nm was immediately increased. This result suggests that the $\mathbf{U^{OH}}$–Gd^{III}–$\mathbf{U^{OH}}$ complexation was completed within 1 min, which is much faster than the entire process of the SDR. Accordingly, the rate-limiting step for the Gd^{III}-triggered SDR is a branch migration step after the metal complexation. Figure 2.20 shows a plausible mechanism for the Gd^{III}-triggered SDR. The addition of Gd^{III} ions immediately triggers the binding of strand **2** to strand **1** through the $\mathbf{U^{OH}}$–Gd^{III}–$\mathbf{U^{OH}}$ complexation. Then the branch migration proceeds to complete the SDR. As suggested by CD spectroscopy shown in Fig. 2.7b, the $\mathbf{U^{OH}}$–Gd^{III}–$\mathbf{U^{OH}}$ complexation resulted in the distortion of the DNA duplex structures. It is highly possible that the branch migration step became slower due to the structural distortion of DNA duplexes caused by the $\mathbf{U^{OH}}$–Gd^{III}–$\mathbf{U^{OH}}$ complexation (Fig. 2.20a). In addition, undesired assemblies through $\mathbf{U^{OH}}$–Gd^{III}–$\mathbf{U^{OH}}$ base pairing in a parallel orientation (Fig. 2.20b) and homodimerization of $\mathbf{U^{OH}}$-containing strands (Fig. 2.20c) may diminish the speed of the Gd^{III}-triggered SDRs.

The reverse SDR was also examined by removing Gd^{III} ions from the duplex with $\mathbf{U^{OH}}$–Gd^{III}–$\mathbf{U^{OH}}$ pairs to regenerate the other duplex containing $\mathbf{U^{OH}}$–A pairs. To duplex $\mathbf{1_F}\cdot\mathbf{2_D}\cdot Gd^{III}{}_4$, 1 equiv. of strand **3** and 4 equiv. of EDTA were added. The emission intensity was gradually increased, indicating the progress of the reverse SDR (Fig. 2.21, blue solid line). However, the speed of the SDR was much slower than that of the Gd^{III}-triggered SDR, even with additional G and C bases at the termini (Fig. 2.21, blue dotted line).

To identify the reason for the slow progress of the EDTA-triggered SDRs, the speed of demetallation was examined by time-course UV–vis absorption measurement (Fig. 2.22). The addition of 4 equiv. of EDTA to 17-mer duplex $\mathbf{4'}\cdot\mathbf{5'}\cdot Gd^{III}{}_4$ resulted in a gradual decrease in the absorbance at 310 nm. This result shows the regeneration of metal-free $\mathbf{U^{OH}}$ nucleobases, confirming that the Gd^{III} ions were removed from

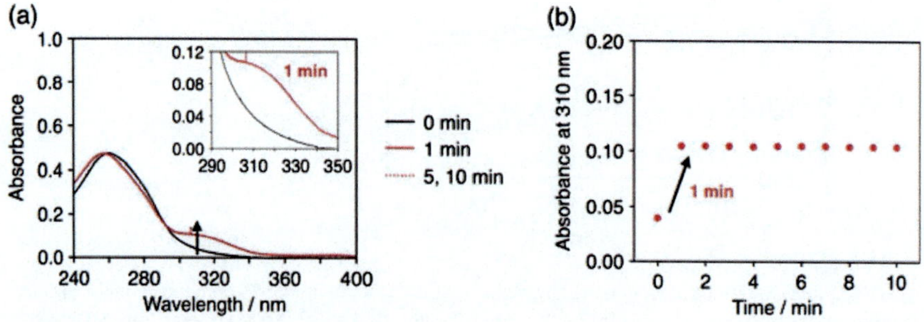

Fig. 2.19 Speed of the $\mathbf{U^{OH}}$–Gd^{III}–$\mathbf{U^{OH}}$ complexation within DNA duplexes. **a** Gd^{III}-responsive change in UV absorption spectra of duplex **4′·5′** containing $\mathbf{U^{OH}}$–$\mathbf{U^{OH}}$ pairs. **b** Time-course plot of the absorbance at 310 nm. [duplex] = 2.0 μM, [Gd^{III}]/[duplex] = 0 or 4 in 10 mM HEPES–NaOH buffer (pH 8.0), 100 mM NaCl. To the preannealed duplex, 4 equiv. of Gd^{III} ions were added. 25 °C, l = 1 cm. The spectra were recorded every 1 min

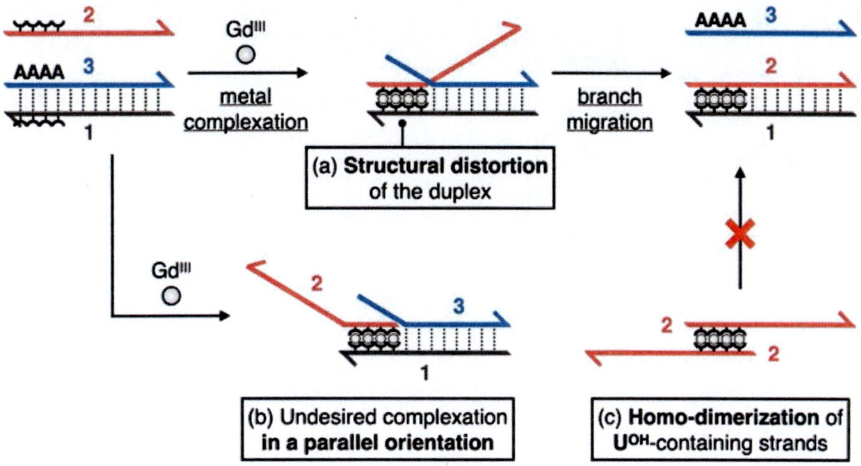

Fig. 2.20 A plausible mechanism of the Gd^{III}-triggered strand displacement reactions (SDRs) using $\mathbf{U^{OH}}$-containing DNA strands

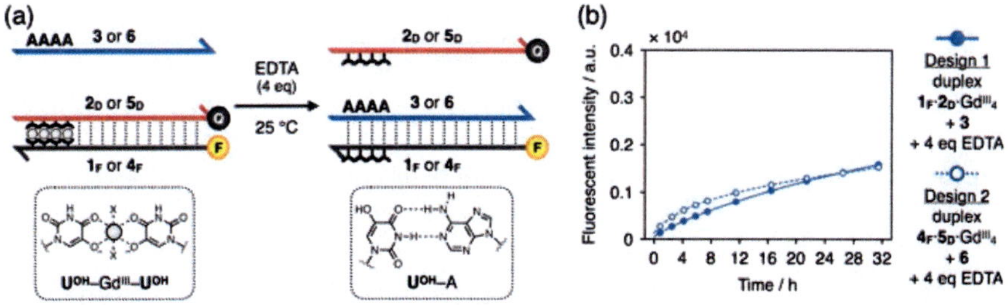

Fig. 2.21 EDTA-triggered reverse strand displacement reactions (SDRs). **a** Scheme. "X" denotes additional ligands coordinating to the Gd^{III} ion such as H_2O and neighboring nucleobases. F: FAM, Q: Dabcyl. **b** Time-course fluorescence analysis of the SDRs. [each DNA] = 2.0 μM, [Gd^{III}]/[each DNA] = 4, [EDTA]/[each DNA] = 4 in 10 mM HEPES–NaOH buffer (pH 8.0), 100 mM NaCl. To the preannealed Gd^{III}-containing duplexes, 1 equiv. of the input strands and 4 equiv. of EDTA were added. The samples were incubated at 25 °C. λ_{ex} = 495 nm, λ_{em} = 519 nm. The fluorescence intensity was recorded every 10 min

duplex **4′·5′** by EDTA. Since the absorbance change was almost converged within 30 min, it was concluded that the demetallation by EDTA is much faster than the whole process of the EDTA-triggered SDR. Therefore, the rate-limiting step for the EDTA-triggered SDR can be the binding of the A-containing strand (**3** or **6**) to the $\mathbf{U^{OH}}$-containing strand ($\mathbf{1_F}$ or $\mathbf{4_F}$) or the subsequent branch migration step.

Furthermore, the SDR was performed by the addition of strand **6** to duplex $\mathbf{4_F \cdot 5_D}$ without Gd^{III} ions (Fig. 2.23, black dotted line). In this case, the rate of the SDR was just comparable with the EDTA-triggered SDR (Fig. 2.23b, blue solid line). This result also suggests that the demetallation step was not the rate-limiting step. In contrast, when the $\mathbf{U^{OH}}$ bases were replaced by natural T bases, the SDR smoothly

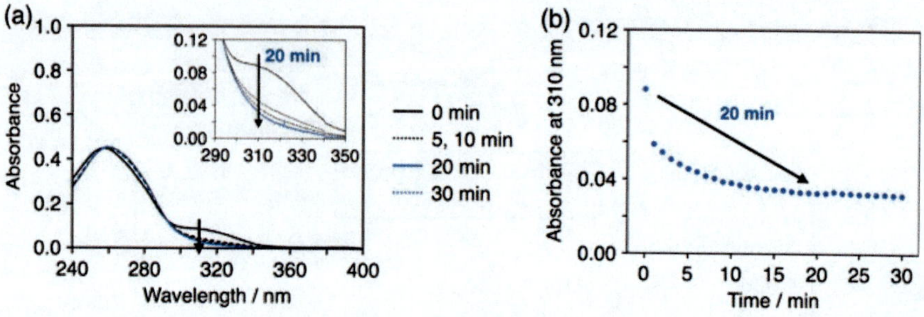

Fig. 2.22 Demetallation of the $\mathbf{U^{OH}}$–Gd^{III}–$\mathbf{U^{OH}}$ complexes within DNA duplexes by the addition of EDTA. **a** UV spectral changes of duplex **4′**·**5′**·Gd^{III}_4 upon addition of EDTA. **b** Time-course plot of the absorbance at 310 nm. [duplex] = 2.0 μM, [Gd^{III}]/[duplex] = 4, [EDTA]/[duplex] = 0 or 4 in 10 mM HEPES–NaOH buffer (pH 8.0), 100 mM NaCl. To the preannealed duplex with 4 equiv. of Gd^{III} ions, 4 equiv. of EDTA were added. 25 °C, l = 1 cm. The spectra were recorded every 1 min

proceeded (Fig. 2.23b, black solid line). This significant acceleration of the SDR can be because the T–A pairs are thermally more stable than $\mathbf{U^{OH}}$–A pairs [25], and thus strand **6** more preferentially invaded through the T–A pairing. Therefore, it appears that the rate of the EDTA-triggered SDR could still be improved by introducing additional natural nucleobases into the $\mathbf{U^{OH}}$-containing terminal sequences, to further increase the binding ability of the toehold region.

In summary, the Gd^{III}-triggered SDRs were demonstrated using $\mathbf{U^{OH}}$-containing DNA strands. Inspired by toehold-mediated SDRs, four $\mathbf{U^{OH}}$ bases were incorporated into the termini of the DNA strands. The base pair switching from the $\mathbf{U^{OH}}$–A pairs to the $\mathbf{U^{OH}}$–Gd^{III}–$\mathbf{U^{OH}}$ pairs at the termini of DNA strands contributed to the efficient SDRs in response to Gd^{III} ions. The reverse SDRs also proceeded by removing Gd^{III} ions with EDTA, albeit slowly.

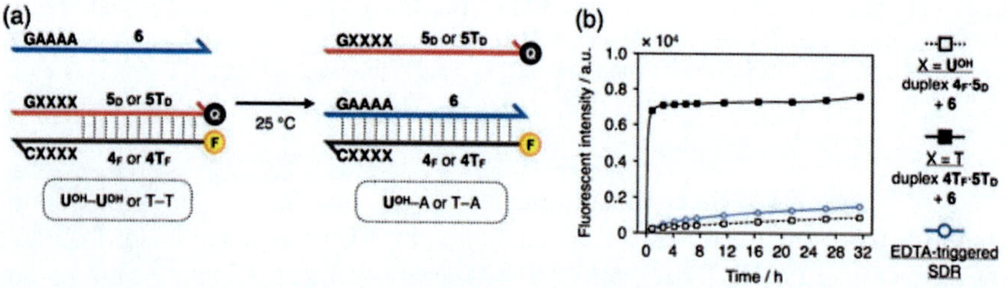

Fig. 2.23 Strand displacement reactions (SDRs) through the invasion of strand **6** into duplexes containing $\mathbf{U^{OH}}$–$\mathbf{U^{OH}}$ or T–T mismatches in the absence of Gd^{III} ions. **a** Scheme. X = $\mathbf{U^{OH}}$ or T. F: FAM, Q: Dabcyl. **b** Time-course fluorescence analysis of the SDRs. [duplex] = 2.0 μM in 10 mM HEPES–NaOH buffer (pH 8.0), 100 mM NaCl. To the preannealed duplex, 1 equiv. of strand **6** was added and the sample was incubated at 25 °C. λ_{ex} = 495 nm, λ_{em} = 519 nm. The fluorescence intensity was recorded every 10 min

2.5 Discussion

I have developed Gd^{III}-responsive DNA strand displacement reactions (SDRs) driven by base pair switching between **U^OH**–A and **U^OH**–Gd^{III}–**U^OH** pairs. The base pair switching of **U^OH** was demonstrated using 16-mer DNA strands containing four **U^OH** bases at the termini. Duplex melting analysis revealed that duplex **1′·2′** with four **U^OH**–**U^OH** mismatch pairs was significantly stabilized in the presence of Gd^{III} ions due to the formation of metal-mediated **U^OH**–Gd^{III}–**U^OH** pairs ($\Delta T_m = +\,26.1$ °C). In contrast, duplex **1′·3′** containing four **U^OH**–A pairs was slightly destabilized by the addition of Gd^{III} ($\Delta T_m = -\,6.4$ °C). As a result, duplex **1′·2′** became much more stable than duplex **1′·3′** in the presence of Gd^{III} ions while duplex **1′·3′** showed higher stability under Gd^{III}-free conditions. Accordingly, it was confirmed that the base-pairing partner of **U^OH** can be switched in a Gd^{III}-responsive manner. In fact, the hybridization preference of **U^OH**-containing strand **1′** was controlled by the addition and removal of Gd^{III} ions.

Furthermore, metal-responsive SDRs were developed based on the base pair switching of **U^OH** (Fig. 2.24). By the addition of Gd^{III} ions to a mixture of duplex $\mathbf{1_F}$·**3** containing **U^OH**–A pairs (25 bp) and **U^OH**-containing invader strand $\mathbf{2_D}$ (25 mer), the SDR proceeded to give duplex $\mathbf{1_F}$·$\mathbf{2_D}$ with **U^OH**–Gd^{III}–**U^OH** pairs. The rate of the Gd^{III}-triggered SDR was slower compared with that of the toehold-mediated SDR using natural strands. Since the **U^OH**–Gd^{III}–**U^OH** complexation was fast enough, it was suggested that the branch migration became slower due to the structural distortion caused by metal complexation. By the addition of EDTA, the reverse SDR to original duplex $\mathbf{1_F}$·**3** with **U^OH**–A pairs also proceeded in spite of its slow kinetics. Thus, it was demonstrated that the base pair switching of **U^OH** is applicable to Gd^{III}-responsive reversible SDRs under isothermal conditions.

SDRs have been an essential tool in dynamic DNA nanotechnology. The introduction of a terminal single-stranded region, "toehold", into DNA strands enabled conventional SDRs through the invasion of input strands and subsequent branch migration. On the other hand, this study features Gd^{III}-responsive SDRs driven by the base pair switching of **U^OH** instead of the "toehold binding". The **U^OH**-containing strand can invade the original duplex through the Gd^{III}-triggered destabilization of **U^OH**–A pairs and formation of **U^OH**–Gd^{III}–**U^OH** pairs at the termini of DNA duplexes. To the best of my knowledge, this toehold-less SDR through branch migration is a totally new concept. Therefore, this study would establish a new design strategy of stimuli-responsive DNA systems.

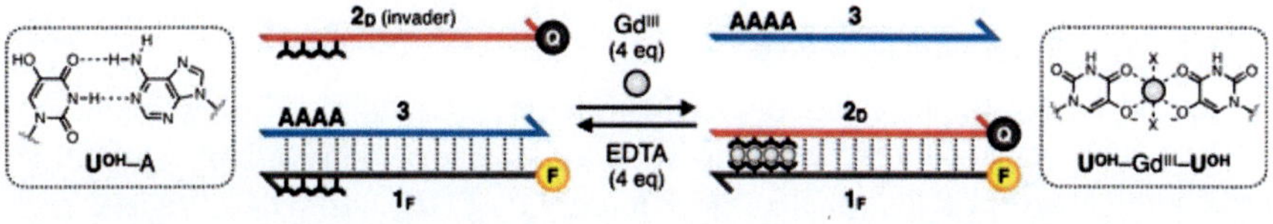

Fig. 2.24 Gd^{III}-responsive DNA strand displacement reactions (SDRs) driven by the base pair switching of the **U^OH** bases

As shown in Sect. 2.1, there have already been several reports on metal-responsive SDRs [23, 24]. In the case of the SDRs triggered by T–Hg^{II}–T base pairing [23], unspecific binding of Hg^{II} ions to the T bases may limit the applicability of the SDRs. In contrast, Gd^{III} ions interact selectively with **U^OH^** bases and do not affect the stability and structures of other regions with natural sequences. Therefore, the Gd^{III}-triggered SDRs are expected to be useful even for controlling higher-ordered DNA nanostructures, which consist of many DNA strands. In addition, the type of input metal ions has been limited since the T–Hg^{II}–T and C–Ag^{I}–C pairs are only available motifs with natural nucleobases. In contrast, other metal ions can be utilized as an input for the SDRs using modified uracil or pyrimidine bases with different functionalities at the 5 positions. In fact, our group has already demonstrated that a 5-carboxyuracil (**caU**) nucleobase shows selective complexation with Cu^{II} ions to form metal-mediated **caU**–Cu^{II}–**caU** base pairs [29]. Therefore, the Gd^{III}-responsive SDR shown in this study will be further expanded to the rational design of SDRs which respond to specific target metal ions.

The SDR triggered by the complexation of 8-hydroxyquinoline (8HQ) and Cu^{II} ions at the termini of DNA duplexes has also been reported [24]. In this previous study, the authors introduced completely synthetic 8HQ ligands into the termini of DNA strands as exposed Cu^{II}-binding sites. Although the Cu^{II}-responsive SDRs smoothly proceeded, this artificial modification may affect the structures and stability of DNA assemblies. On the other hand, a modified uracil base, $\mathbf{U^{OH}}$, was introduced as a metal-responsive unit in this study. Since $\mathbf{U^{OH}}$ retains the Watson–Crick face of the natural uracil or thymine base, it is highly compatible with natural DNA structures. In addition, both the original and product duplexes behaved as a full-match duplex through the $\mathbf{U^{OH}}$–A and $\mathbf{U^{OH}}$–Gd^{III}–$\mathbf{U^{OH}}$ base pairing, respectively. Therefore, this Gd^{III}-responsive SDR using $\mathbf{U^{OH}}$ is highly suitable for switching two distinct states, each of which can be stabilized through the two different base pairing modes. Furthermore, the Gd^{III}-triggered SDR is reversible by the addition of EDTA to remove Gd^{III} ions and restore the original $\mathbf{U^{OH}}$–A base pairing. The "bifacial" base-pairing feature of $\mathbf{U^{OH}}$ would be important for further applying the Gd^{III}-triggered SDR to dynamic switching of DNA assemblies. The application of the base pair switching of $\mathbf{U^{OH}}$ to metal-responsive DNA nanodevices will be discussed in Chap. 4.

2.6 Experimental Section

General

All natural DNA strands including FAM-modified strands and Dabcyl-modified strands were purchased from Japan Bio Service Co., Ltd. (Saitama, Japan) at HPLC purification grade. T4 Polynucleotide Kinase (T4 PNK), T4 DNA Ligase, and NEBNext® Quick Ligation Reaction Buffer were purchased from New England Biolabs, Inc. DNA strands containing 5-hydroxyuracil ($\mathbf{U^{OH}}$) nucleobases were synthesized on an Applied Biosystems 394 DNA synthesizer on a 1-μmol scale as described below. Some of the **U^OH^**-containing strands were prepared by enzymatic ligation reactions using shorter **U^OH^**-containing strands. The synthesized DNA strands were purified by reverse-phase HPLC (Waters XBridge C18 column) or by denaturing polyacrylamide gel electrophoresis. All oligonucleotides were characterized by matrix-assisted laser desorption ionization-time-of-flight (MALDI-TOF) or electrospray ionization-time-of-flight (ESI-TOF) mass spectrometry. MALDI-TOF mass spectra were recorded on Bruker ultrafleXtreme and Autoflex III using a mixture of 3-hydroxypicolinic acid (3-HPA) and ammonium citrate as a matrix. ESI-TOF mass spectra were measured by a Waters Micromass LCT premier. DNA concentrations were calculated based on UV absorbance at 260 nm. The molar extinction coefficients (ε_{260}) of the strands were calculated based on the nearest-neighbor method ($\varepsilon_{260} = 4.59 \times 10^3$ M^{-1} cm^{-1} for the $\mathbf{U^{OH}}$ nucleoside). FAM- or Dabcyl-labeled strands were quantified based on UV absorbance at 495 nm and 476 nm ($\varepsilon_{495} = 7.50 \times 10^3$ M^{-1} cm^{-1} for FAM, $\varepsilon_{476} = 3.20 \times 10^3$ M^{-1} cm^{-1} for Dabcyl), respectively. Metal sources were purchased from FUJIFILM Wako Pure Chemical Industries ($MnCl_2{\cdot}4H_2O$ (99.9% purity), $FeCl_2{\cdot}4H_2O$ (99.9%), $CoCl_2{\cdot}6H_2O$ (99.5%), $NiCl_2{\cdot}6H_2O$ (99.9%), and $CuCl_2{\cdot}2H_2O$ (99.9%)) and Soekawa Chemical Co. ($ZnSO_4{\cdot}7H_2O$ (99.9%), $GdCl_3{\cdot}6H_2O$ (99.9%), and $EuCl_3{\cdot}6H_2O$ (99.9%)).

Solid-Phase DNA Synthesis

$\mathbf{U^{OH}}$ phosphoramidite was synthesized from 2′-deoxyuridine in four steps according to the reported protocols [28, 29] and immediately used for solid-phase DNA synthesis. DNA strands containing 5-hydroxyuracil ($\mathbf{U^{OH}}$) nucleobases were synthesized on an Applied Biosystems 394 DNA synthesizer on a 1-μmol scale in a DMTr-on mode with ultramild deprotection phosphoramidites and reagents (Glen Research). DNA synthesis was performed according to the standard procedure except for an extended coupling time for the $\mathbf{U^{OH}}$ nucleoside (15 min). The synthesized strands were cleaved from the solid support and deprotected using a 25% NH_3 aqueous solution (1.0 mL) at room temperature for 4 h. The obtained DNA strands were first detritylated and purified by a Poly-Pak II cartridge (Glen Research). After further purification by reverse-phase HPLC (Waters XBridge Oligonucleotide BEH C18 column, buffers: A = MeCN, B = 0.1 M TEAA (pH 7.0) + 2% MeCN, 60 °C), all DNA strands were identified by ESI-TOF mass spectrometry.

Strand **1′**. 5′ - CAC ATT GTT GTA $\mathbf{U^{OH}U^{OH}U^{OH}}$ $\mathbf{U^{OH}}$ - 3′. HPLC retention time: 27.7 min (gradient: 4% A (0 min), 6.5% A (40 min)). ESI MS: *m*/*z* calcd for $[C_{154}H_{194}N_{49}O_{104}P_{15} - 7H]^{7-}$: 693.1, found: 693.2.

Strand **2′**. 5′ - $\mathbf{U^{OH}U^{OH}U^{OH}}$ $\mathbf{U^{OH}}$TA CAA CAA TGT G - 3′. HPLC retention time: 23.7 min (gradient: 4% A (0 min), 6.5% A (40 min)). ESI MS: *m*/*z* calcd for $[C_{154}H_{192}N_{55}O_{100}P_{15} - 7H]^{7-}$: 695.7, found: 695.8.

Strand **4′**. 5′ - CAC ATT GTT GTA $\mathbf{U^{OH}U^{OH}U^{OH}}$ $\mathbf{U^{OH}}$C - 3′. HPLC retention time: 32.5 min (gradient: 4% A (0 min), 6.5% A (40 min)). ESI MS: *m*/*z* calcd for $[C_{163}H_{206}N_{52}O_{110}P_{16} - 5H]^{5-}$: 1028.8, found: 1028.8.

Strand **5′**. 5′ - G**U^{OH}U^{OH} U^{OH}U^{OH}**T ACA ACA ATG TG - 3′. HPLC retention time: 32.0 min (gradient: 4% A (0 min), 6.5% A (40 min)). ESI MS: *m*/*z* calcd for $[C_{164}H_{204}N_{60}O_{106}P_{16} - 4H]^{4-}$: 1300.7, found: 1300.9.

Synthesis of U^{OH}-Containing DNA Strands by Enzymatic Ligation Reactions

For the synthesis of strands **1** and **4** with four **U^{OH}** bases at the 3′ termini, the 5′ termini of the shorter strands **1′** and **4′** were phosphorylated by T4 PNK. Each strand was diluted to be 40 μM in NEBNext® Quick Ligation Reaction Buffer. After the addition of T4 PNK (0.8 U/μL), the reaction solution was incubated at 37 °C for 1 h. After heated at 95 °C for 10 min, the obtained donor strand was mixed with FAM-labeled primer strands and template strands. After annealing the mixture (85 °C → 4 °C, 1 °C/min), T4 DNA Ligase was added to the solution. The final concentrations of the reaction solutions were 10 μM donor, 12 μM primer, 12 μM template, and 1 U/μL T4 DNA Ligase. The reaction was conducted at 25 °C for 1 h. Equal volume of 50 mM EDTA aq was added to quench the reaction and the reaction solution was heated at 95 °C for 10 min.

Strands **2** and **5** possessing **U^{OH}** bases at the 5′ termini were synthesized using strands **2′** and **5′** as primer strands in the enzymatic ligation reactions. After the addition of phosphorylated donor strands and template strands, the ligation reactions by T4 DNA Ligase were performed as described above.

The obtained mixture was purified by isopropanol precipitation, and the desired products were separated from the template strands by denaturing PAGE. The isolated products were further purified by size-exclusion chromatography (Sephadex G-25 Fine, GE Healthcare) and the subsequent isopropanol precipitation. All DNA strands were identified by MALDI-TOF mass spectrometry, and the amount of DNA strands was determined based on the UV absorbance value at 260 nm.

Strand **1_F**. 5′ - FAM - AGC TAC CTC CAC ATT GTT GTA **U^{OH}U^{OH}U^{OH} U^{OH}** - 3′. MALDI MS: *m*/*z* calcd for $[C_{267}H_{329}N_{81}O_{167}P_{25} - H]^{-}$: 8117.3, found: 8116.6.

Strand **2_D**. 5′ - **U^{OH}U^{OH}U^{OH} U^{OH}**TA CAA CAA TGT GGA GGT GAC T – Dabcyl - 3′. MALDI MS: *m*/*z* calcd for $[C_{264}H_{329}N_{81}O_{167}P_{25} - H]^{-}$: 8180.4, found: 8179.2.

Strand **4_F**. 5′ - AGC TAC CTC CAC ATT GTT GTA **U^{OH}U^{OH}U^{OH} U^{OH}**C - 3′. MALDI MS: *m*/*z* calcd for $[C_{276}H_{340}N_{84}O_{173}P_{26} - H]^{-}$: 8400.3, found: 8401.3.

Strand **5_D**. 5′ - G**U^{OH}U^{OH} U^{OH}U^{OH}**T ACA ACA ATG TGG AGG TGA CT - Dabcyl - 3′. MALDI MS: *m*/*z* calcd for $[C_{274}H_{341}N_{101}O_{166}P_{26} - H]^{-}$: 8505.5, found: 8505.2.

Duplex Melting Analysis

All samples were prepared by mixing DNA strands (2.0 μM) in 10 mM HEPES–NaOH buffer (pH 8.0) with 100 mM NaCl. After the addition of Gd^{III} ions (0–6 equiv), the sample solutions were annealed (from 85 °C to 4 °C, − 1.0 °C/min). Absorbance at 260 nm (A_{260}) was monitored by UV-1800 and UV-1900 spectrophotometers (Shimadzu) equipped with a TMSPC-8 temperature controller. The temperature was raised from 4 to 85 °C at a rate of 0.2 °C/min. The melting curves were drawn after normalizing the absorbance value at 260 nm (A_{260}) as follows:

$$\text{Normalized } \Delta A_{260} = \{A_{260}(t\,^{\circ}\text{C}) - A_{260}(4\,^{\circ}\text{C})\} / \{A_{260}(85\,^{\circ}\text{C}) - A_{260}(4\,^{\circ}\text{C})\}.$$

The melting temperature (T_m) was calculated from the inflection point of the melting curve using the T_m analysis software LabSolutions (Shimadzu) with a 17-point adaptive smoothing program. The T_m values were determined from the average of at least three independent runs.

Metal Complexation of Duplexes Containing U^{OH}–U^{OH} Pairs

UV–vis absorption spectra were recorded on a V-730 spectrophotometer (JASCO) at 4 °C with a path length of 1 cm. The samples (2.0 μM) were prepared in HEPES–NaOH buffer (pH 8.0) containing 100 mM NaCl. The sample solutions were annealed (from 85 °C to 4 °C, – 1.0 °C/min) in the presence of varying concentrations of Gd^{III} ions (0–6 equiv.) prior to the measurement.

Circular dichroism (CD) spectra were measured by a JASCO J-820 spectropolarimeter with 10-time accumulation using a path length of 1 cm at 4 °C. The samples were prepared as described above. The obtained spectra were smoothed using a simple moving average smoothing program.

ESI-TOF mass spectra were measured by a Waters Micromass LCT premier. The samples (40 μM) were prepared in 20 mM NH_4OAc buffer (pH 7.0) and annealed before the measurements (from 85 °C to 4 °C, – 1.0 °C/min) in the presence of Gd^{III} ions (4 equiv.). The measurement was conducted in a negative mode.

Metal-Dependent Control of the Hybridization Preference of U^{OH}-Containing DNA Strands

For quantification of the hybridization products, strands **3′** and **6′** were labeled with a FAM. The sample solutions were prepared by mixing the DNA strands (2.0 μM each) in 10 mM HEPES–NaOH buffer (pH 8.0) with 100 mM NaCl. After the addition of Gd^{III} (0–8 equiv.), the solutions were annealed (85 °C → 4 °C, – 1.0 °C/min). All samples were analyzed by native gel electrophoresis with 20% polyacrylamide gel in a cool incubator (4 °C). Each band was quantified based on FAM fluorescence, and the yield was determined by comparing the band intensity with that of the authentic samples. The averages of three or more independent experiments were calculated as the yields.

Metal-Responsive Strand Exchange Using U^{OH}-Containing Strands

The Gd^{III}-responsive strand exchange under isothermal conditions was performed as follows. The sample solutions were prepared by mixing the DNA strands (2.0 μM each) in 10 mM HEPES–NaOH buffer (pH 8.0) with 100 mM NaCl and annealed (85 °C → 4 °C, – 1.0 °C/min). After the addition of Gd^{III} ions (4 equiv.), the samples were incubated at 25 °C. Certain amount of the solution was taken out at defined time points and kept frozen. All the samples were analyzed by native gel electrophoresis as shown above.

For EDTA-triggered strand exchange, the sample solutions were prepared by mixing the DNA strands (2.0 μM each) and Gd^{III} (8.0 μM) in 10 mM HEPES–NaOH buffer (pH 8.0) with 100 mM NaCl and annealed (85 °C → 4 °C, – 1.0 °C/min).

After EDTA (4 equiv.) was added, the samples were incubated at 25 °C. Samples taken at defined time points were analyzed by native gel electrophoresis as described above.

For reversible switching of the hybridization partner, the sample solutions were prepared by mixing the DNA strands (2.0 μM each) in 10 mM HEPES–NaOH buffer (pH 8.0) with 100 mM NaCl and annealed (85 °C → 4 °C, − 1.0 °C/min). Gd^{III} (4 equiv.) and EDTA (4 equiv.) were alternately added at 2-h intervals at 25 °C. Samples taken at defined time points were analyzed by native gel electrophoresis as described above.

Gd^{III}-Triggered DNA Strand Displacement Reactions (SDRs)

Fluorescence intensity was recorded on a JASCO FP-8350 fluorometer with a path length of 0.5 cm. For time-course analysis of the SDRs by fluorescence measurement, strands $\mathbf{1_F}$ and $\mathbf{4_F}$ were labeled with a fluorophore (FAM) and stands $\mathbf{2_D}$ and $\mathbf{5_D}$ with a quencher (Dabcyl). The original duplexes ($\mathbf{1_F \cdot 3}$ or $\mathbf{4_F \cdot 6}$) were prepared by mixing the two strands (2.0 μM each) in 10 mM HEPES–NaOH buffer (pH 8.0) with 100 mM NaCl. The samples were annealed (85 °C → 25 °C, − 1.0 °C/min) and then incubated at 25 °C. After the addition of the other strand ($\mathbf{2_D}$ or $\mathbf{5_D}$), the SDR was initiated by the addition of Gd^{III} ions (4 equiv.). The progress of the SDR at 25 °C was monitored based on the changes in the emission intensity of the FAM ($\lambda_{ex} = 495$ nm, $\lambda_{em} = 519$ nm). The fluorescent intensity was measured in 10-min intervals for 8 h and subsequently in 30-min intervals. In the control experiments using natural T bases instead of $\mathbf{U^{OH}}$, strand $\mathbf{5T_D}$ and Gd^{III} (4 equiv.) were added to the pre-annealed duplex $\mathbf{4T_F \cdot 6}$. For the toehold-mediated SDRs, strand $\mathbf{6_D}$ was added to duplex $\mathbf{4T_F \cdot 5s}$.

Metal Selectivity Study of the SDRs Using U^{OH}-Containing DNA Strands

Fluorescence intensity was recorded on a Varioskan LUX plate-reader (Thermo Fisher). Original duplex $\mathbf{1_F \cdot 3}$ was prepared by mixing the two strands (2.0 μM each) in 10 mM HEPES–NaOH buffer (pH 8.0) with 100 mM NaCl. The samples were annealed (85 °C → 25 °C, − 1.0 °C/min) and then incubated at 25 °C. After the addition of strand $\mathbf{2_D}$, the SDR was initiated by the addition of metal ions (4 equiv.). The progress of the SDR at 25 °C was monitored based on the changes in the emission intensity of the FAM ($\lambda_{ex} = 495$ nm, $\lambda_{em} = 520$ nm). The fluorescent intensity was measured in 5-min intervals for 20 h.

EDTA-Triggered Reverse SDRs

Fluorescence intensity was recorded on a JASCO FP-8350 fluorometer with a path length of 0.5 cm. The original duplexes ($\mathbf{1_F \cdot 2_D}$·$Gd^{III}{}_4$ or $\mathbf{4_F \cdot 5_D}$·$Gd^{III}{}_4$) were prepared by mixing the two strands (2.0 μM each) and Gd^{III} ions (4 equiv.) in 10 mM HEPES–NaOH buffer (pH 8.0) with 100 mM NaCl. The samples were annealed (85 °C → 25 °C, − 1.0 °C/min) and then incubated at 25 °C. After the addition of the other strand (**3** or **6**), the SDR was initiated by the addition of EDTA (4 equiv.). The progress of the SDR at 25 °C was monitored based on the changes in the emission intensity of the FAM ($\lambda_{ex} = 495$ nm, $\lambda_{em} = 519$ nm). The fluorescent intensity was measured in 10-min intervals for 8 h and subsequently in 30-min intervals. In the control experiments without Gd^{III} ions, strand **6** was just added to duplex $\mathbf{4_F{\cdot}5_D}$ or $\mathbf{4T_F{\cdot}5T_D}$.

Evaluation of the Rate of Metal Complexation

UV–vis absorption spectra were recorded on a V-730 spectrophotometer (JASCO) at 25 °C with a path length of 1 cm. The samples (2.0 μM) were prepared in HEPES–NaOH buffer (pH 8.0) containing 100 mM NaCl. For Gd^{III}-triggered complexation, the sample solutions were annealed (from 85 °C to 25 °C, − 1.0 °C/min) and then mixed with Gd^{III} ions (4 equiv.). For EDTA-triggered demetallation, the samples were annealed in the presence of Gd^{III} ions (4 equiv.) and then mixed with EDTA (4 equiv.). The UV spectrum was recorded in 1-min intervals.

References

1. Zhang DY, Seelig G (2011) Nat Chem 3:103–113
2. Simmel FC, Yurke B, Singh HR (2019) Chem Rev 119:6326–6369
3. Yurke B, Mills AP Jr (2013) Genet Progr Evol Mach 4:111–122
4. Zhang DY, Winfree E (2009) J Am Chem Soc 131:17303–17314
5. Zhang DY, Turberfield AJ, Yurke B, Winfree E (2007) Science 318:1121–1125
6. Qian L, Winfree E, Bruck J (2011) Nature 475:368–372
7. Srinivas N, Parkin J, Seelig G, Winfree E, Soloveichik D (2017) Science 358:eaal2052
8. Morihiro K, Kodama T, Waki R, Obika S (2014) Chem Sci 5:744–750
9. Kou B, Zhang J, Huai X, Liang X, Xiao SJ (2015) RSC Adv 5:5055–5058
10. Cheng B, Kashida H, Shimada N, Maruyama A, Asanuma H (2017) ChemBioChem 18:1568–1572
11. Amodio A, Zhao B, Porchetta A, Idili A, Castronovo M, Fan C, Ricci F (2014) J Am Chem Soc 136:16469–16472
12. Kankanamalage DVDW, Tran JHT, Beltrami N, Meng K, Zhou X, Pathak P, Isaacs L, Burin AL, Ali MF, Jayawickramarajah J (2022) J Am Chem Soc 144:16502–16511
13. Wang C, O'Hagan MP, Li Z, Zhang J, Ma X, Tian H, Willner I (2022) Chem Soc Rev 51:720–760
14. Lubbe AS, Szymanski W, Feringa BL (2017) Chem Soc Rev 46:1052–1079
15. Mattath MN, Ghosh D, Pratihar S, Shi S, Govindaraju T (2022) ACS Omega 7:3167–3176
16. Kamiya Y, Asanuma H (2014) Acc Chem Res 47:1663–1672
17. Lohmann F, Ackermann D, Famulok M (2012) J Am Chem Soc 134:11884–11887
18. Sillner EM, Kamada Y, Suzuki Y, Emura T, Hidaka K, Dietz H, Sugiyama H, Endo M (2017) Angew Chem Int Ed 129:15526–15530
19. Yang Y, Endo M, Hidaka K, Sugiyama H (2012) J Am Chem Soc 134:20645–20653
20. Green L, Amodio A, Subramanian HKK, Ricci F, Franco E (2017) Nano Lett 17:7283–7288
21. Peng L, You M, Yuand Q, Wu C, Han D, Chen Y, Zhong Z, Xue J, Tan W (2012) J Am Chem Soc 134:12302–12307
22. Guo W, Lu CH, Orbach R, Wang F, Qi XJ, Cecconello A, Seliktar D, Willner I (2015) Adv Mater 27:73–78
23. Ding W, Deng W, Zhu H, Liang H (2013) Chem Commun 49:9953–9955
24. Wang L-L, Zhang Q-L, Wang Y, Liu Y, Lin J, Xie F, Xu L (2021) Chem Sci 12:8698–8705
25. Takezawa Y, Nishiyama K, Mashima T, Katahira M, Shionoya M (2015) Chem Eur J 21:14713–14716
26. Nishiyama K, Takezawa Y, Shionoya M (2016) Inorg Chim Acta 452:176–180
27. Morningstar ML, Kreutzer DA, Essigmann JM (1997) Chem Res Toxicol 10:1345–1350
28. Fujimoto J, Tran L, Sowers LC (1997) Chem Res Toxicol 10:1254–1258

29. Takezawa Y, Suzuki A, Nakaya M, Nishiyama K, Shionoya M (2020) J Am Chem Soc 142:21640–21644

Chapter 3
Metal-Dependent Base Pair Switching of *N*,*N*-Dicarboxymethyl-5-Aminouracil Nucleosides

Abstract The base pair switching of $\mathbf{U^{OH}}$ has had limited application because it requires the consecutive introduction of multiple $\mathbf{U^{OH}}$ bases into DNA strands. In this chapter, I aimed to develop metal-responsive base pair switching by a singly incorporated ligand-type nucleobase. To this end, I designed *N*,*N*-dicarboxymethyl-5-aminouracil (**dcaU**) containing an iminodiacetate ligand. The **dcaU** base was expected to show higher coordination ability than $\mathbf{U^{OH}}$ due to its negative charges and chelate effects. After obtaining **dcaU** phosphoramidite, **dcaU**-containing DNA strands were synthesized by an automated DNA synthesizer. In the presence of Gd^{III} ions, a DNA duplex with a **dcaU**–**dcaU** pair was significantly stabilized due to the formation of **dcaU**–Gd^{III}–**dcaU** pairs. As a result, the addition of Gd^{III} ions reversed the relative stability of two duplexes containing **dcaU**–A or **dcaU**–**dcaU** pairs, indicating base pair switching between **dcaU**–A and **dcaU**–Gd^{III}–**dcaU**. The base pair switching of **dcaU** was further applicable for Gd^{III}-responsive DNA strand exchange. These results showed that the efficiency of the base pair switching can be rationally controlled by careful design of ligand-type nucleobases.

Keywords Base pair switching · Modified uracil · Iminodiacetic acid · Chelate effect · Solid-phase DNA synthesis

3.1 Introduction

Modified uracil bases ($\mathbf{U^X}$) with a metal-binding ligand at the 5 positions have been developed as a "bifacial" nucleobase. Conceptually, $\mathbf{U^X}$ bases can form hydrogen-bonded $\mathbf{U^X}$–A base pairs and metal-mediated $\mathbf{U^X}$–M–$\mathbf{U^X}$ base pairs in the absence and presence of certain metal ions (M), respectively. Our group previously reported bifacial base-pairing behaviors of 5-hydroxyuracil ($\mathbf{U^{OH}}$) and 5-carboxyuracil (**caU**) bases [1–4]. DNA duplexes with $\mathbf{U^X}$–$\mathbf{U^X}$ pairs can be thermally stabilized by the formation of metal-mediated base pairs such as $\mathbf{U^{OH}}$–Gd^{III}–$\mathbf{U^{OH}}$ and **caU**–Cu^{II}–**caU** pairs. On the other hand, duplexes containing $\mathbf{U^X}$–A pairs were destabilized by the addition of the corresponding metal ions. Accordingly, metal-responsive conversion

K. Mori, *Metal-Responsive Base Pair Switching of Ligand-type Uracil Nucleobases*, Springer Theses, https://doi.org/10.1007/978-981-99-9400-7_3

of the base-pairing modes, termed as "base pair switching", would be useful for dynamic control of DNA assemblies.

In Chap. 2, metal-responsive base pair switching of the **U^OH** bases between hydrogen-bonded **U^OH**–A pairs and metal-mediated **U^OH**–Gd^{III}–**U^OH** pairs is described. Since DNA strand displacement reactions (SDRs) were successfully driven by the metal-triggered base pair switching, it was expected to be further applicable to metal-responsive control of DNA assemblies. However, it has been required to consecutively introduce multiple **U^OH**–**U^OH** pairs into DNA duplexes to construct **U^OH**–Gd^{III}–**U^OH** base pairs. This is probably because the formation of a single **U^OH**–Gd^{III}–**U^OH** complex cannot compensate for duplex destabilization resulting from structural distortion accompanied by metal complexation. This limitation of the sequence design prevents further application of the 5-modified uracil bases to metal-responsive DNA systems.

In this study, I aimed to develop a novel 5-modified uracil base which forms a single metal-mediated base pair within DNA duplexes. For this purpose, an iminodiacetic acid (IDA) ligand was introduced at the 5 position of the uracil base and an *N*,*N*-dicarboxymethyl-5-aminouracil (**dcaU**) nucleobase was designed (Fig. 3.1a). IDA is a versatile tridentate ligand which forms stable complexes with various transition metal ions and lanthanide ions [5–11], and is a substructure of many chelating agents such as EDTA and BAPTA (Fig. 3.1b) [12]. The **dcaU** base was expected to behave as a tetradentate ligand by coordinating the IDA and 4-carbonyl group, forming **dcaU**–M–**dcaU** (M = metal ions) complexes with an octacoordinate structure (Fig. 3.1a). Since lanthanide ions generally prefer a high coordination number [13, 14], Gd^{III} was chosen as the central metal ion for the formation of a metal-mediated **dcaU**–M–**dcaU** base pair in DNA duplexes. The chelating effect and negative charges of the IDA ligand would endow the **dcaU** base with enhanced coordination ability to Gd^{III} ions compared to the **U^OH** bases. Furthermore, the **dcaU** base retains the hydrogen-bonding site of natural uracil or thymine bases, and thus forms a hydrogen-bonded **dcaU**–A pair (Fig. 3.1a). Therefore, the **dcaU** base was expected to behave as a bifacial nucleobase exhibiting Gd^{III}-responsive base pair switching between **dcaU**–A and **dcaU**–Gd^{III}–**dcaU** pairs.

In this chapter, I demonstrate metal-responsive base pair switching of the **dcaU** base and its application to reversible control of DNA hybridization. First, I have established synthetic protocols of **dcaU** nucleosides and **dcaU**-containing DNA strands (Sect. 3.2). Then metal complexation of the **dcaU** nucleoside was examined at a monomeric level, confirming 2:1 complexation with Gd^{III} ions (Sect. 3.3). Furthermore, base-pairing behaviors of the **dcaU** base within DNA duplexes were studied. It was revealed that **dcaU** can form both a metal-mediated **dcaU**–Gd^{III}–**dcaU** pair and a hydrogen-bonded **dcaU**–A pair by incorporation of a single **dcaU** base into DNA sequences (Sects. 3.4 and 3.5). Finally, the bifacial base-pairing behavior of **dcaU** was utilized for reversibly switching hybridization preferences of the **dcaU**-containing DNA strands in a Gd^{III}-responsive manner (Sect. 3.6).

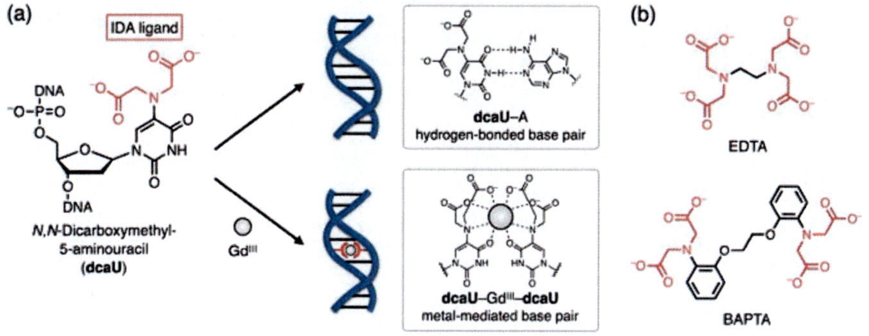

Fig. 3.1 Design of a novel *N*,*N*-dicarboxymethyl-5-aminouracil (**dcaU**) nucleobase. **a** Chemical structure and Gd^{III}-dependent bifacial base-pairing behavior of the **dcaU** base. **b** Chemical structures of chelating agents containing the iminodiacetic acid (IDA) ligand

3.2 Synthesis of DNA Strands Containing *N*,*N*-Dicarboxymethyl-5-Aminouracil Bases

3.2.1 Overview

I have synthesized a **dcaU** nucleoside and incorporated it into DNA strands. The synthetic scheme was designed as shown in Fig. 3.2. To introduce an IDA ligand onto the 5 position of the uracil nucleobase, 5-bromo-2′-deoxyuridine **1** was used as a starting material. After converting the 5-bromo group to an amino group [15], methyl bromoacetate was introduced to obtain methyl-protected **dcaU** nucleoside **3**. Subsequent deprotection of nucleoside **3** with NaOH gave the desired **dcaU** nucleoside **4** as a disodium salt. To incorporate **dcaU** nucleosides into DNA strands, the 5′-OH group of nucleoside **3** was first protected by a DMTr group. Then the 3′-OH was functionalized with phosphoramidite to obtain nucleoside **6**, which was immediately used for solid-phase synthesis of **dcaU**-containing DNA strands by an automated synthesizer. The resulting DNA strands were treated with NaOH for cleavage from the solid support and deprotection, and finally purified by reverse-phase HPLC.

3.2.2 Synthesis of Methyl-Protected dcaU *Nucleoside 3* (Fig. 3.3)

5-Amino-2′-deoxyuridine **2** was synthesized from 5-bromo-2′-deoxyuridine **1** according to the previous study [15]. After the nucleophilic substitution of the 5-bromo group by benzylamine (Fig. 3.4a), hydrogenation with Pd/C under a H_2 atmosphere was conducted to obtain nucleoside **2** (Fig. 3.4b).

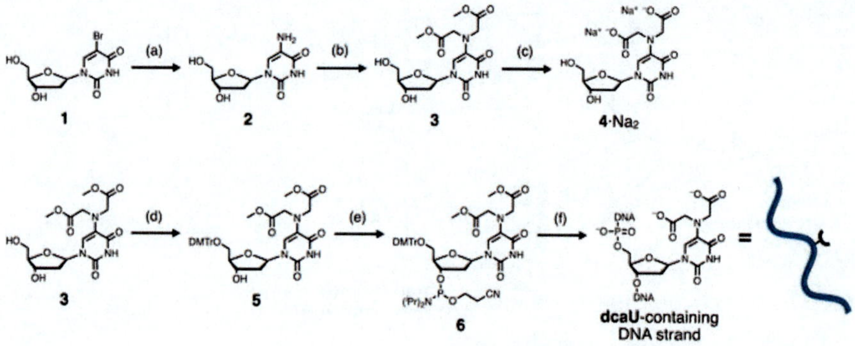

Fig. 3.2 Synthetic route for **dcaU** nucleoside **4** and **dcaU**-containing DNA strands. (a) (i) Benzylamine, 90 °C; (ii) H_2, Pd/C, MeOH, 25 °C. (b) Methyl bromoacetate, 2,6-lutidine, MeOH, 60 °C, 38% in 2 steps. (c) 0.1 M NaOH aq, 60 °C, quant. (d) DMTrCl, DMAP, pyridine, rt, 59%. (e) $NC(CH_2)_2OP[N(^iPr)_2]_2$, diisopropylamine, 1*H*-tetrazole, CH_2Cl_2, rt, 65%. (f) Solid-phase DNA synthesis by an automated synthesizer. Synthesized DNA strands were cleaved from the solid support and deprotected by 0.3 M NaOH aq at 37 °C. DMTr = 4,4′-dimethoxytrityl, DMAP = 4-dimethylaminopyridine, ^iPr = isopropyl

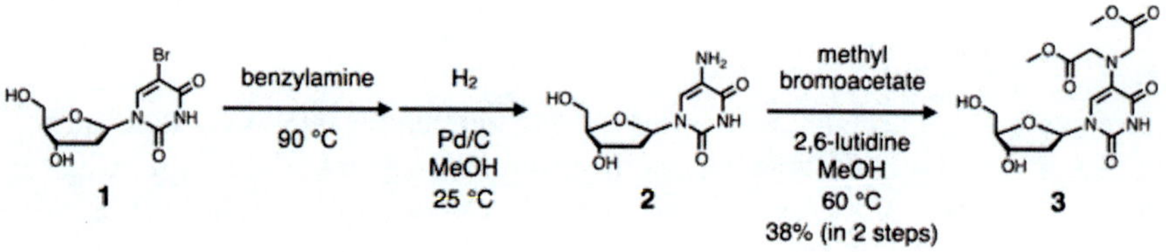

Fig. 3.3 Synthesis of methyl-protected **dcaU** nucleoside **3**

The crude product of **2** was directly reacted with methyl bromoacetate in methanol to introduce two carboxymethyl moieties by nucleophilic attack of the 5-amino group. First, the reaction was performed at 60 °C in the presence of pyridine as a base. However, the reaction did not completely proceed even after adding excess amount of methyl bromoacetate, resulting in the selective formation of monosubstituted nucleoside **3′**. This can be because the reactivity of the 5-amino group was diminished after the first substitution reaction and methyl bromoacetate was consumed by undesired reactions with pyridine. In fact, ^1H NMR spectra of the reaction mixture showed that the crude mixture contained nucleoside **3′** and the pyridine-derived byproduct (Fig. 3.5). Nucleoside **3′** was isolated by silica gel chromatography and identified by ^1H NMR spectroscopy (Fig. 3.6).

To prevent unnecessary reactions that consume methyl bromoacetate, 2,6-lutidine was used as a base instead of pyridine. It was expected that the nucleophilicity of the aromatic nitrogen of 2,6-lutidine is lower than that of pyridine due to the steric hindrance of the 2- and 6-methyl groups while it retains enough basicity to promote the reaction. The crude product of nucleoside **2** was reacted with methyl bromoacetate in methanol at 60 °C in the presence of 2,6-lutidine. After the addition

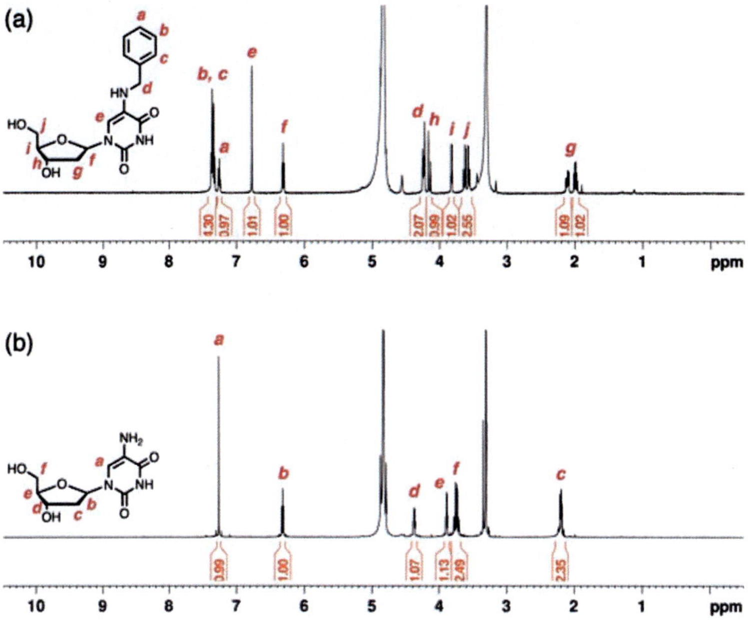

Fig. 3.4 ^{1}H NMR spectra of **a** 5-benzylamino-2′-deoxyuridine and **b** 5-amino-2′-deoxyuridine **2** after the reactions. 500 MHz, 300 K, CD_3OD

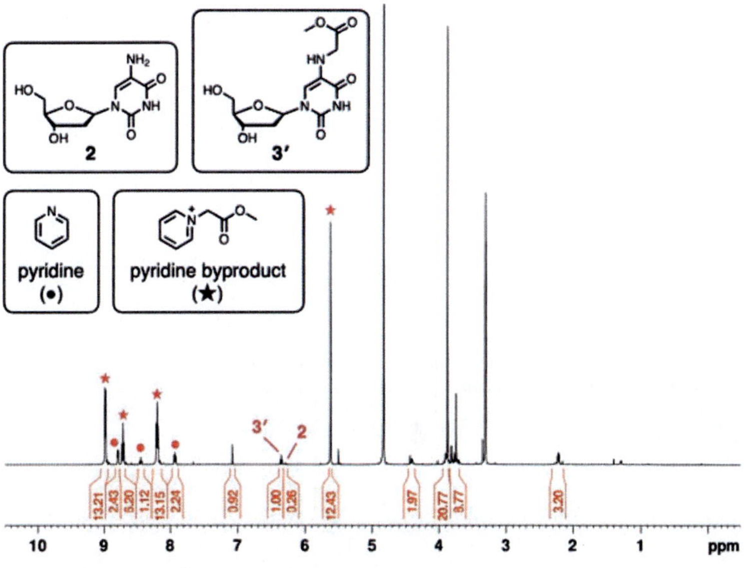

Fig. 3.5 ^{1}H NMR spectrum of the crude mixture after the reaction to synthesize nucleoside **3** in the presence of pyridine as a base. 500 MHz, 300 K, CD_3OD

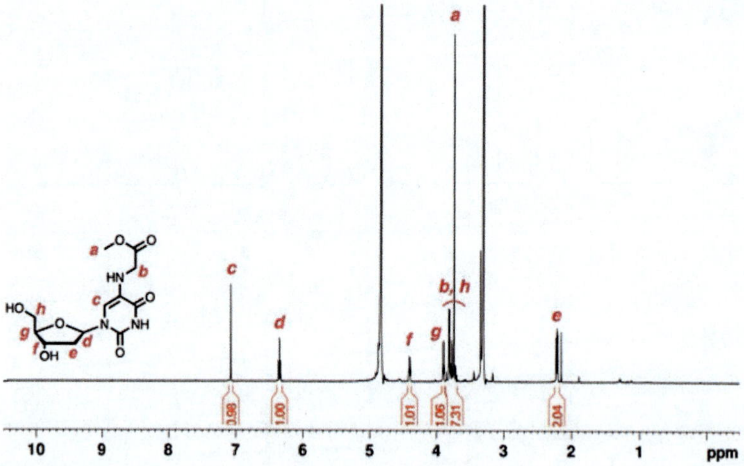

Fig. 3.6 ^{1}H NMR spectrum of nucleoside **3′** after purified by silica gel chromatography. 500 MHz, 300 K, CD_3OD

of excess methyl bromoacetate, the ^{1}H NMR spectrum of the reaction mixture showed a signal at 4.02 ppm corresponding to four protons of the two carboxymethyl groups, indicating the formation of the desired disubstituted compound **3** (Fig. 3.7). The crude mixture was purified by silica gel column chromatography and pure target compound **3** was obtained in 38% yield in two steps (Fig. 3.8). Nucleoside **3** was also identified by ESI mass spectrometry (see Sect. 3.8).

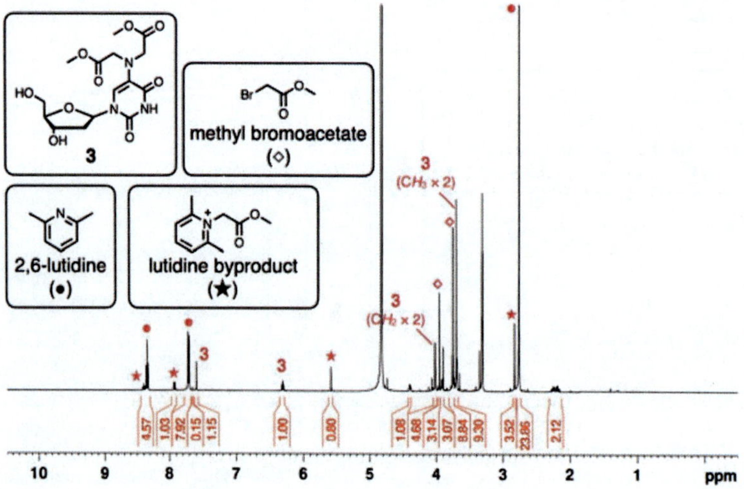

Fig. 3.7 ^{1}H NMR spectrum of the crude mixture after the reaction to synthesize nucleoside **3** in the presence of 2,6-lutidine as a base. 500 MHz, 300 K, CD_3OD

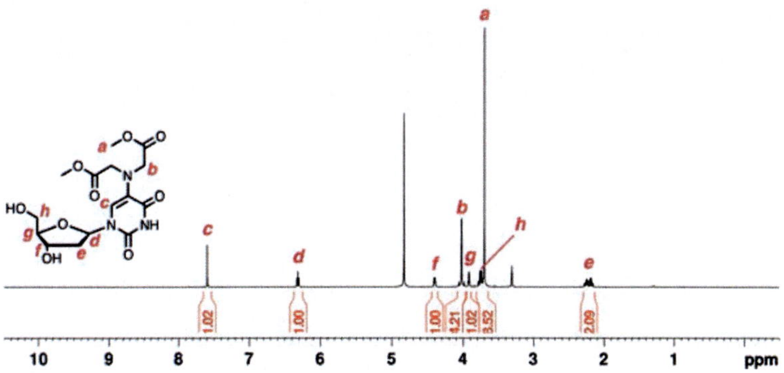

Fig. 3.8 ^{1}H NMR spectrum of methyl-protected **dcaU** nucleoside **3** after purified by silica gel chromatography. 500 MHz, 300 K, CD_3OD

3.2.3 Synthesis of dcaU *Nucleoside 4* (Fig. 3.9)

dcaU nucleoside **4** was synthesized by deprotection of nucleoside **3** under basic conditions. To obtain disodium salt of nucleoside **4**, methyl-protected nucleoside **3** was dissolved in 2 equiv. of NaOH aqueous solution and incubated at room temperature for 4 h. ^{1}H NMR spectrum of the reaction solution indicated that the reaction gave mainly two species, possibly fully deprotected nucleoside **4** and partially deprotected byproduct **4′**, in almost a 1:1 ratio (Fig. 3.10).

To enhance the reaction progress, the reaction conditions were modified by raising the reaction temperature to 60 °C and elongating the incubation time to 22 h. ^{1}H NMR spectrum of the obtained product did not show any signals of the methyl protecting group, indicating that the desired nucleoside **4** was quantitatively formed as a disodium salt (Fig. 3.11). The formation of the target nucleoside **4**·Na_2 was also confirmed by ESI mass spectrometry (see Sect. 3.8).

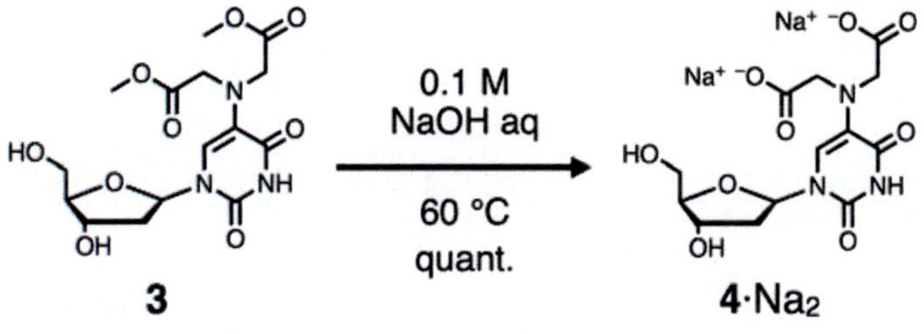

Fig. 3.9 Synthesis of a disodium salt of **dcaU** nucleoside **4**

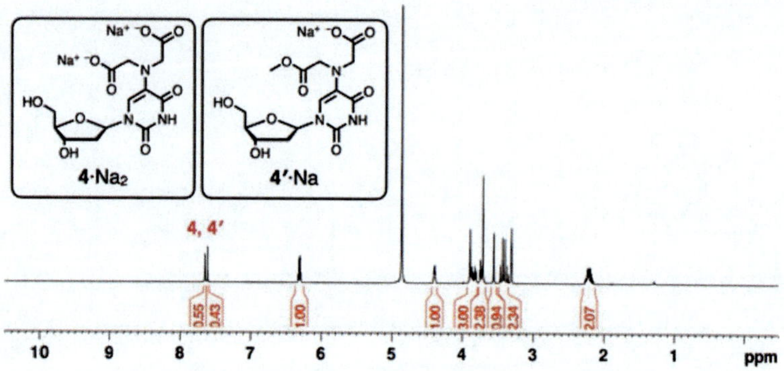

Fig. 3.10 ^{1}H NMR spectrum of the reaction solution to synthesize **dcaU** nucleoside **4** after treating nucleoside **3** by NaOH aq at room temperature for 4 h. 500 MHz, 300 K, CD_3OD

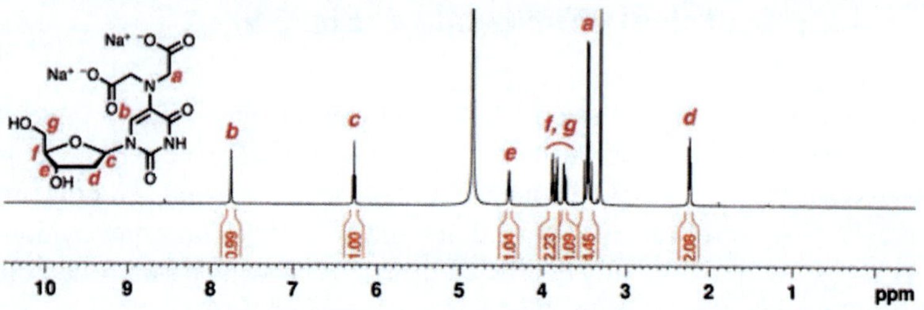

Fig. 3.11 ^{1}H NMR spectrum of **dcaU** nucleoside **4**·Na_2 obtained by treating nucleoside **3** with NaOH aq at 60 °C for 22 h. 500 MHz, 300 K, CD_3OD

3.2.4 Synthesis of DMTr-Protected dcaU *Nucleoside 5* (Fig. 3.12)

For solid-phase synthesis of DNA strands by an automated synthesizer, the 5′-OH group must be protected with a DMTr group to prevent undesired reactions. The DMTr group is removed under acidic conditions during the deprotection step of

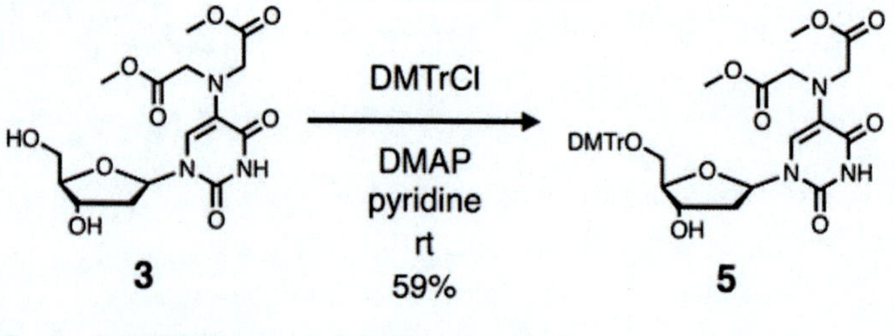

Fig. 3.12 Synthesis of DMTr-protected **dcaU** nucleoside **5**

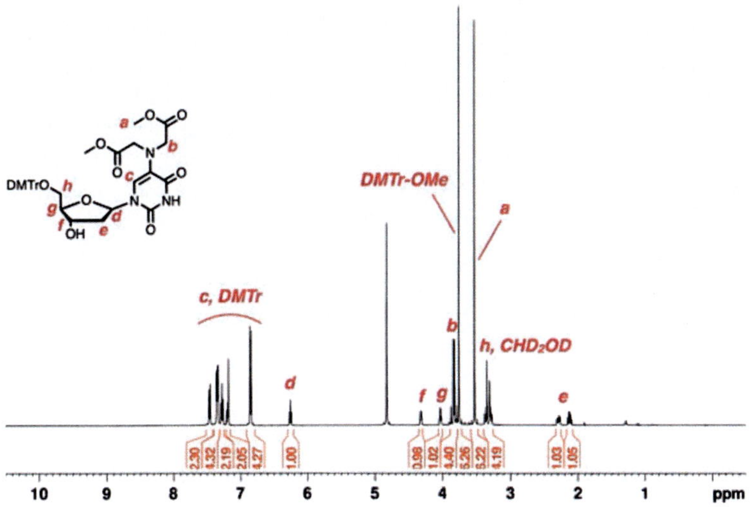

Fig. 3.13 ^{1}H NMR spectrum of DMTr-protected **dcaU** nucleoside **5**. 500 MHz, 300 K, CD_3OD

DNA synthesis, and the coupling yield can be calculated from the UV absorbance of the free DMTr species.

Methyl-protected **dcaU** nucleoside **4** was mixed with DMTrCl and DMAP in pyridine, and the reaction was conducted at room temperature for 2 h. After purification by silica gel column chromatography, the target compound **5** was obtained in 59% yield (Fig. 3.13).

3.2.5 *Synthesis of* dcaU *Phosphoramidite 6* (Fig. 3.14)

Finally, the 3′-OH group was functionalized by phosphoramidite to synthesize compound **6** as a precursor of solid-phase DNA synthesis. Nucleoside **5** was reacted with 2-cyanoethyl-*N,N,N',N'*-tetraisopropylphosphorodiamidite in the presence of

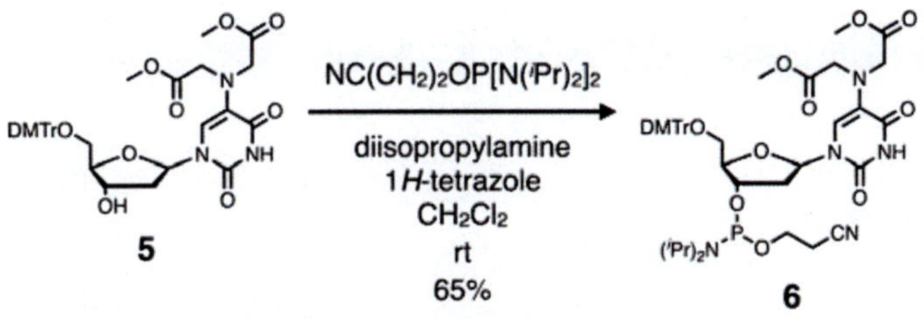

Fig. 3.14 Synthesis of **dcaU** phosphoramidite **6**

diisopropylammonium tetrazolide in dichloromethane for 75 min. Subsequent purification by silica gel column chromatography gave a diastereomeric mixture of the target compound **6** in 65% yield. ^{1}H NMR spectrum of the obtained product indicated the formation of two diastereomers (Fig. 3.15a). In the ^{31}P NMR spectrum, two signals corresponding to the diastereomers were observed at 149.22 and 148.80 ppm, which are in the region characteristic of phosphoramidite derivatives (Fig. 3.15b). Since phosphoramidite compounds are generally unstable under ambient conditions, the resulting nucleoside **6** was immediately used for solid-phase DNA synthesis. It should be noted that the diastereomeric mixture can be used since the phosphoramidite is oxidized to an achiral phosphodiester during DNA synthesis.

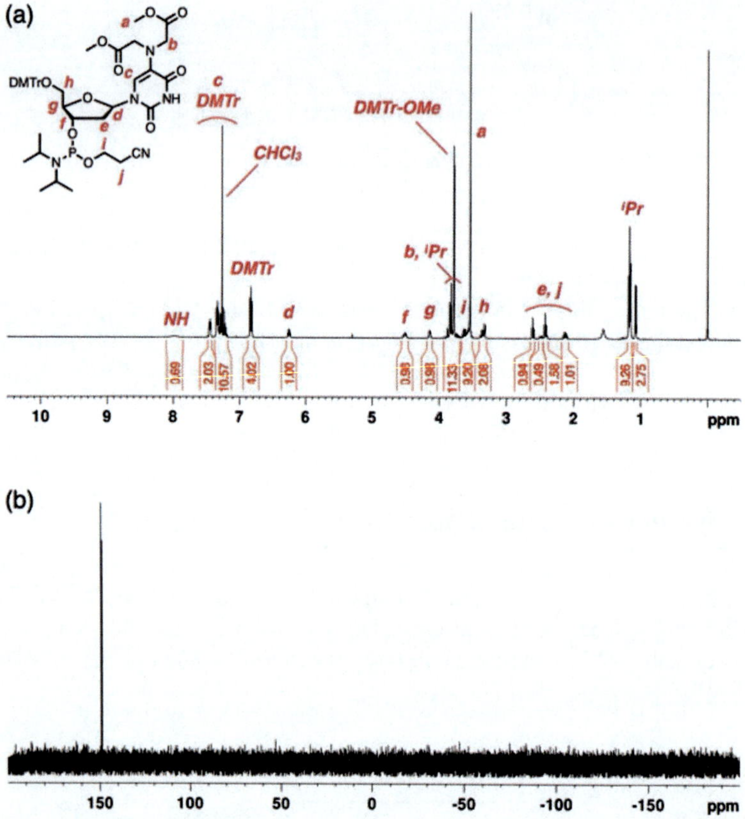

Fig. 3.15 **a** ^{1}H NMR and **b** ^{31}P NMR spectra of **dcaU** phosphoramidite **6**. **a** 500 MHz, **b** 202 MHz, 300 K, $CDCl_3$

Table 3.1 Sequences of DNA strands used in Chap. 3

Name	Sequence (5′ to 3′)
7	CAC ATT A**dcaU**T GTT GTA
8	TAC AAC A**dcaU**T AAT GTG
7T	CAC ATT A**T**T GTT GTA
8T	TAC AAC A**T**T AAT GTG
8N (N = A, T, G, or C)	TAC AAC A**N**T AAT GTG
8A′	TAC AAC A**A**T AAT G

3.2.6 *Synthesis of DNA Strands Containing the* dcaU *Base*

To investigate base-pairing behaviors of the **dcaU** base inside DNA duplexes, 15-mer DNA strands **7** and **8** containing one **dcaU** base in the middle were designed (Table 3.1). Since strands **7** and **8** have complementary sequences with each other except for the **dcaU** bases, a metal-mediated **dcaU**–M–**dcaU** pair can be constructed inside duplex **7**·**8** in the presence of appropriate metal ions. Duplex **7T**·**8T** with one T–T mismatch was also prepared for control experiments. The **dcaU** base was also expected to form a hydrogen-bonded base pair with a natural adenine (A). The formation of the **dcaU**–A pair was studied using duplexes **7**·**8N** containing a **dcaU**–N pair (N = A, T, G, or C).

DNA strands **7** and **8** containing one **dcaU** base in the middle were synthesized by well-established solid-phase DNA synthesis using an automated synthesizer. The **dcaU** phosphoramidite synthesized in Sect. 3.2.5 was immediately used for DNA synthesis. The coupling reaction of the **dcaU** nucleotide was performed with an extended reaction time (15 min) to improve the coupling yield. Then the obtained DNA strands were treated by 0.3 M NaOH aq for cleavage from the solid support and removal of the protecting groups on the nucleobases and the phosphate backbones. Subsequent reverse-phase HPLC purification gave the desired strands **7** and **8** with high purity (Fig. 3.16). The DNA strands were characterized by ESI mass spectrometry (Fig. 3.17, Table 3.2).

In conclusion, I have synthesized **dcaU** nucleosides and **dcaU**-containing DNA strands. **dcaU** nucleoside **4** and **dcaU** phosphoramidite **6** were synthesized from 5-bromo-2′-deoxyuridine **1** in good yields. The **dcaU**-containing DNA strands were obtained using **dcaU** phosphoramidite **6** for standard solid-phase synthesis. Since the obtained products were all characterized well and showed high purity, they were further used for metal complexation experiments.

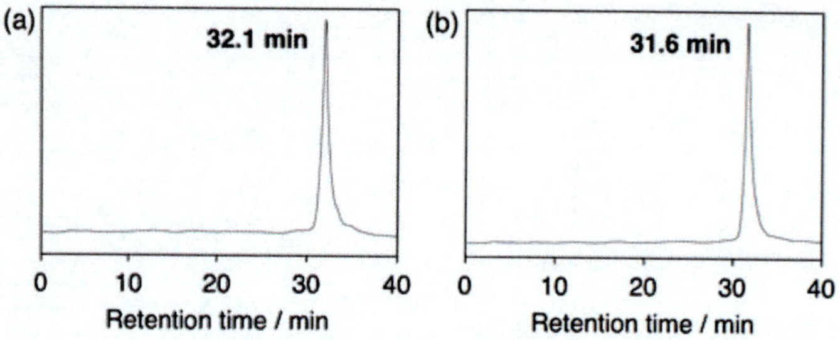

Fig. 3.16 HPLC analysis of DNA strands containing **dcaU** nucleotides after purification. **a** Strand **7**, **b** strand **8**. Waters XBridge Oligonucleotide OST C18 column (4.6 × 50 mm), flow rate: 1.18 mL/min, temperature: 60 °C, monitored at 260 nm. Gradient: 0 to 40 min, 4% A to 6.5% A (solvent A = MeCN, solvent B = 0.1 M TEAA buffer (pH 7.0) + 2% MeCN)

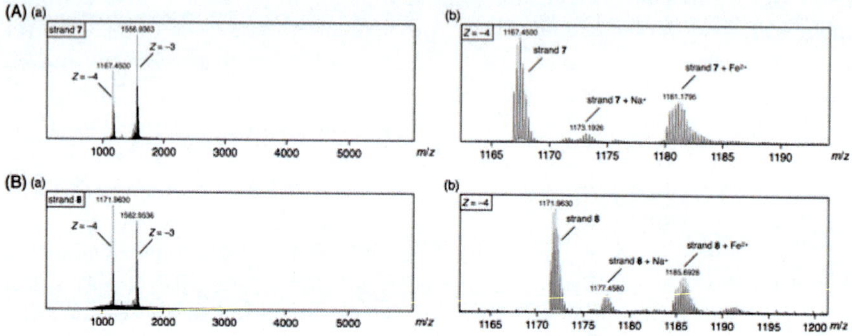

Fig. 3.17 ESI mass spectra of DNA strands containing **dcaU** nucleotides after purification. **A** Strand **7**, **B** strand **8**. **a** Whole spectra and **b** enlarged spectra are shown. [DNA] = 5.0 μM in H_2O, eluted with MeOH/20 mM NH_4OAc aq (1/1), negative mode

Table 3.2 Characterization of **dcaU**-containing DNA strands by ESI mass spectrometry

Strand	Sequence (5′ to 3′)	Formula	Calculated mass	Observed mass
7	CAC ATT Aca**U**T GTT GTA	$[C_{151}H_{191}N_{51}O_{95}P_{14} - 4H]^{4-}$	1167.51	1167.45
8	TAC AAC Aca**U**T AAT GTG	$[C_{151}H_{189}N_{57}O_{91}P_{14} - 4H]^{4-}$	1171.70	1171.96

3.3 Metal Complexation of *N*,*N*-Dicarboxymethyl-5-Aminouracil Nucleosides

With a view to applying the **dcaU** base to metal-mediated base pairing within DNA duplexes, I investigated the formation of a 2:1 complex of **dcaU** nucleosides and a metal ion (Fig. 3.18a). The **dcaU** base was expected to function as a tetradentate ligand through coordination of the IDA ligand and the 4-carbonyl group. Thus, a Gd^{III}

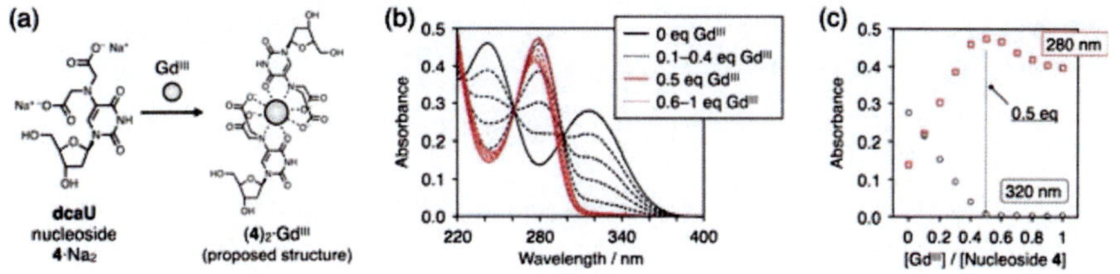

Fig. 3.18 Metal complexation of **dcaU** nucleoside **4**. **a** Scheme. **b** UV–vis absorption spectra of nucleoside **4** in the presence of different concentrations of Gd^{III} ions. [nucleoside **4**] = 200 μM, $[Gd^{III}]$/[**4**] = 0 (black solid line), 0.1, 0.2, 0.3, 0.4 (black dotted lines), 0.5 (red solid line), 0.6, 0.7, 0.8, 0.9, and 1.0 (red dotted lines) in 10 mM HEPES–NaOH buffer (pH 7.0), $l = 0.3$ cm, 25 °C. **c** Plot of the absorbance at 280 and 320 nm against the ratio of $[Gd^{III}]$/[nucleoside **4**]

ion was selected as a bridging metal ion considering that lanthanide ions generally prefer eight or more coordination numbers [13, 14].

Gd^{III} complexation of the **dcaU** nucleoside was first studied by UV–vis absorption spectroscopy. The **dcaU** nucleoside **4** was mixed with varying concentrations of Gd^{III} ions and UV–vis spectra were recorded. In the absence of Gd^{III}, two absorption bands around 240 nm and 320 nm were observed (Fig. 3.18b). Upon addition of Gd^{III} ions, those absorption peaks were gradually decreased and a new absorption band appeared around 280 nm. The absorbance at 280 nm was linearly enhanced by increasing the amount of Gd^{III} ions with isosbestic points at 261 and 297 nm. The spectral change was converged when 0.5 equiv. of Gd^{III} were added (Fig. 3.18c). This result indicated 2:1 complexation between the **dcaU** nucleoside and Gd^{III} ions. The quantitative formation of **dcaU**–Gd^{III}–**dcaU** complexes was further confirmed by Job's plot analysis based on UV absorption changes (Fig. 3.19). Furthermore, a signal of the (**dcaU**)$_2$·Gd^{III} complex was predominantly observed by ESI mass spectrometry (found: 874.07 ($z = +1$); calcd for $[(\mathbf{4})_2 + Gd^{III} + 2H]^+$: 874.10), indicating the selective formation of the expected 2:1 complexes (Fig. 3.20).

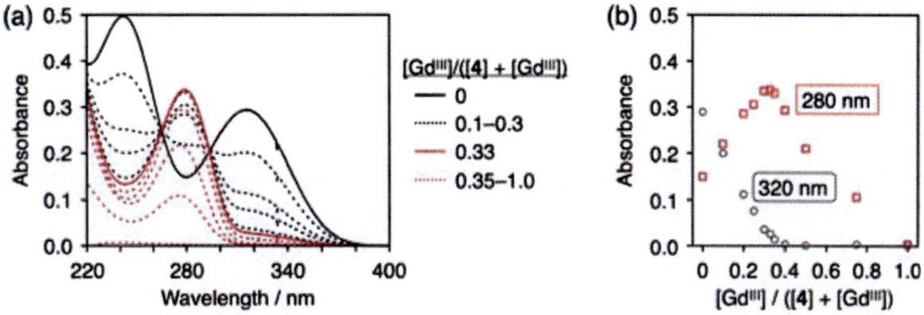

Fig. 3.19 Job's plot analysis of the metal complexation of **dcaU** nucleoside **4**. **a** UV–vis absorption spectra of nucleoside **4** with the varying ratios of nucleoside **4** and Gd^{III} ions. [nucleoside **4**] + $[Gd^{III}]$ = 200 μM, $[Gd^{III}]$/([nucleoside **4**] + $[Gd^{III}]$) = 0 (black solid line), 0.10, 0.20, 0.25, 0.30 (black dotted lines), 0.33 (red solid line), 0.35, 0.40, 0.50, 0.75, and 1.0 (red dotted lines) in 10 mM HEPES–NaOH buffer (pH 7.0), $l = 0.3$ cm, 25 °C. **b** Plots of the absorbance at 280 and 320 nm against the ratios of $[Gd^{III}]$/([**4**] + $[Gd^{III}]$)

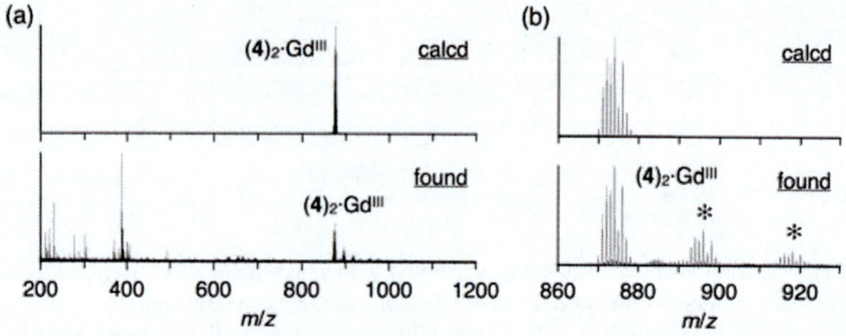

Fig. 3.20 ESI mass spectra of **dcaU** nucleoside **4** with 0.5 equiv. of Gd^{III} ions. **a** Whole spectrum, **b** enlarged spectrum. Small signals (*) were attributed to sodium adducts. [nucleoside **4**] = 100 μM, $[Gd^{III}]/[\mathbf{4}] = 0.50$ in 10 mM NH_4OAc buffer (pH 7.0), positive mode

The metal-dependent UV spectral change of the **dcaU** nucleoside was similar to that of the previously reported BAPTA ligand [12]. BAPTA has been known as a chelating agent showing high affinity to various transition metal ions and lanthanide ions through coordination of two IDA ligands and two ether oxygen atoms. It was reported that the characteristic absorption change of the BAPTA ligand in response to metal ions derives from coordination of the nitrogen atom, which alters the electronic structure of the π-conjugated benzene rings. Accordingly, it was speculated that the **dcaU** base can form octacoordinated 2:1 complexes with Gd^{III} ions through coordination of the IDA ligands and the 4-carbonyl oxygens.

In summary, the 2:1 complexation between **dcaU** nucleosides and a Gd^{III} ion in a monomeric level was confirmed by UV–vis spectroscopy and ESI mass spectrometry. These results encouraged me to further study the **dcaU**–Gd^{III}–**dcaU** complexation within DNA duplexes, namely, metal-mediated **dcaU**–Gd^{III}–**dcaU** base pairing.

3.4 Metal-Mediated Base Pairing of *N*,*N*-Dicarboxymethyl-5-Aminouracil Bases within DNA Duplexes

As described in Sect. 3.3, the **dcaU** nucleoside showed quantitative 2:1 complexation with a Gd^{III} ion at a monomeric level. Encouraged by this result, herein I studied Gd^{III}-mediated base pairing of the **dcaU** base within DNA duplexes. For this purpose, I prepared 15-mer DNA duplex **7**·**8** containing one **dcaU**–**dcaU** pair in the middle as a Gd^{III}-binding site.

First, thermal stability of duplex **7**·**8** was examined in the absence and presence of Gd^{III} ions by duplex melting analysis. In the absence of Gd^{III} ions, melting temperature (T_m) of duplex **7**·**8** was 22.5 °C, which was significantly lower than that of a natural full-match duplex **7T**·**8A** containing a T–A pair instead of the **dcaU**–**dcaU** pair (T_m = 47.9 °C) (Fig. 3.21). This result indicated that the **dcaU**–**dcaU** pair

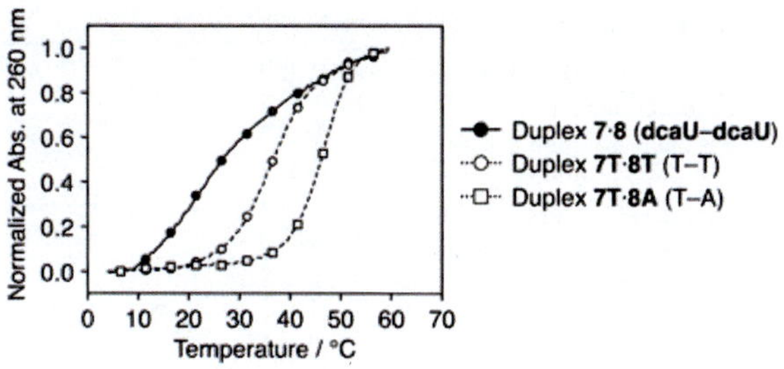

Fig. 3.21 Melting curves of DNA duplexes containing a **dcaU–dcaU** (**7·8**), T–T (**7T·8T**), or T–A (**7T·8A**) pair in the middle. [duplex] = 2.0 μM in 10 mM HEPES–NaOH buffer (pH 7.0), 100 mM NaCl, 0.2 °C/min

behaves as a mismatch in the DNA duplex. Furthermore, the T_m value of duplex **7·8** was lower than that of duplex **7T·8T** containing a T–T mismatch (T_m = 37.4 °C). The low thermal stability of duplex **7·8** can be attributed to the electrostatic repulsion between the negatively charged **dcaU** bases facing to each other. This destabilizing effect is often the case for artificial nucleobases modified with carboxylate ligands such as 5-carboxyuracil (**caU**) [4] and 4-carboxyimidazole (**Im^C^**) [16, 17].

Duplex **7·8** was significantly stabilized (T_m = 38.6 °C, ΔT_m = + 16.1 °C) by the addition of 1 equiv. of Gd^{III} ions (Fig. 3.22a). In contrast, duplex **7·8T** and duplex **7T·8T** containing a **dcaU**–T and a T–T pair instead of the **dcaU–dcaU** pair, respectively, did not show any Gd^{III}-responsive changes in the thermal stability (Fig. 3.22b, c). Therefore, it was concluded that the Gd^{III}-dependent thermal stabilization of duplex **7·8** was caused by the interstrand **dcaU**–Gd^{III}–**dcaU** complexation.

The stoichiometry of the complexation between the **dcaU** base and Gd^{III} ions was further examined by performing duplex melting analysis of duplex **7·8** with varying amounts of Gd^{III} ions (Fig. 3.23). In the presence of 0.5 equiv. of Gd^{III} ions, the melting curve exhibited a two-step transition, suggesting that both a Gd^{III}-free duplex with a **dcaU–dcaU** mismatch and a Gd^{III}-bridged duplex with a **dcaU**–Gd^{III}–**dcaU** pair were formed. When 2 equiv. of Gd^{III} ions were added, the melting curve was shifted to a lower-temperature direction compared to that with 1 equiv. of Gd^{III} ions. This result indicated that duplex **7·8** was destabilized in the presence of excess Gd^{III} ions, possibly because 1:1 complexation between the **dcaU** base and a Gd^{III} ion prevented the formation of the interstrand **dcaU**–Gd^{III}–**dcaU** complexes. Based on

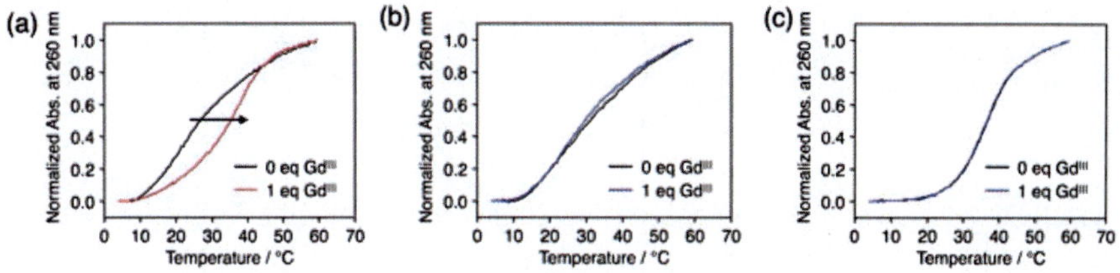

Fig. 3.22 Melting curves of DNA duplexes containing a **a dcaU–dcaU** (**7·8**), **b dcaU**–T (**7·8T**), or **c** T–T pair (**7T·8T**) in the middle in the absence and presence of Gd^{III} ions. [duplex] = 2.0 μM, [Gd^{III}]/[duplex] = 0 (black lines) and 1 (red or purple lines) in 10 mM HEPES–NaOH buffer (pH 7.0), 100 mM NaCl, 0.2 °C/min

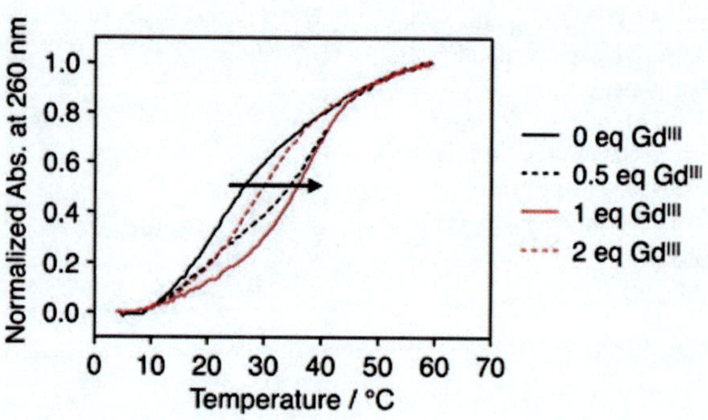

Fig. 3.23 Melting curves of duplex **7·8** containing a **dcaU–dcaU** pair in the presence of varying amounts of Gd^{III} ions. [duplex] = 2.0 μM, [Gd^{III}]/[duplex] = 0 (black solid line), 0.5 (black dotted line), 1 (red solid line), and 2 (red dotted line) in 10 mM HEPES–NaOH buffer (pH 7.0), 100 mM NaCl, 0.2 °C/min

these results, the 2:1 complexation between **dcaU** and Gd^{III} was strongly supported, confirming the quantitative formation of a single metal-mediated **dcaU**–Gd^{III}–**dcaU** pair within DNA duplexes.

The selectivity of the **dcaU** base for metal ions was investigated by duplex melting analysis in the presence of other transition metals such as Fe^{III}, Ni^{II}, Cu^{II}, and Zn^{II}. As shown in Fig. 3.24, only Gd^{III} induced significant thermal stabilization of duplex **7·8** among all the metal ions tested. Considering that the IDA ligand normally forms stable complexes with transition metal ions such as Fe^{III} and Cu^{II} [7–9], this result suggests that the coordination of the 4-carbonyl group of the **dcaU** base also plays a key role for the metal-mediated base pairing. Since Gd^{III} ions prefer high coordination numbers, the tetradentate feature of the **dcaU** base is suitable for the metal-mediated **dcaU**–M–**dcaU** base pairing by 2:1 complexation. The + 3 charge of Gd^{III} ions may also be important for efficient complexation with the **dcaU** base by compensating for the negative charges on **dcaU**.

The effects of the **dcaU**–Gd^{III}–**dcaU** complexation on the duplex structure were studied by circular dichroism (CD) spectroscopy (Fig. 3.25). The CD spectra of duplex **7·8** showed a positive Cotton effect at 279 nm and a negative Cotton effect at 249 nm (Fig. 3.25, red solid line), suggesting that the normal right-handed B-type duplex structure was retained even after introducing the unnatural **dcaU** bases. Compared to natural full-match duplex **7T·8A** (Fig. 3.25, black solid line), duplex

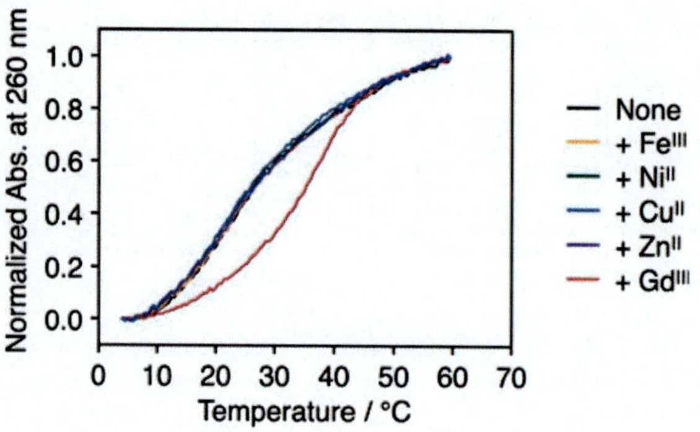

Fig. 3.24 Melting curves of duplex **7·8** containing a **dcaU–dcaU** pair in the presence of various metal ions. [duplex] = 2.0 μM, [metal ion]/[duplex] = 1 in 10 mM HEPES–NaOH buffer (pH 7.0), 100 mM NaCl, 0.2 °C/min

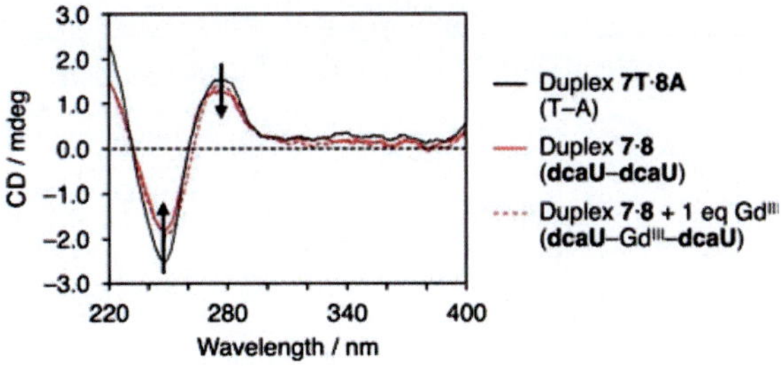

Fig. 3.25 CD spectra of DNA duplexes containing a T–A (**7T·8A**, black solid line), **dcaU–dcaU** (**7·8**, red solid line), or **dcaU**–Gd^{III}–**dcaU** pair (**7·8**·Gd^{III}, red dotted line). [duplex] = 2.0 μM, 10 mM HEPES–NaOH buffer (pH 7.0), 100 mM NaCl, l = 0.3 cm, 5 °C

7·8 exhibited lower CD intensity. This can be because the duplex structure was unwound by introducing the **dcaU–dcaU** pair with a bulky and nonplanar structure. The addition of Gd^{III} ions did not cause a significant change in the CD spectra (Fig. 3.25, red dotted line). These results suggest that the **dcaU**–Gd^{III}–**dcaU** complex was accommodated into the DNA duplex as a metal-mediated base pair without severe distortion of the duplex structure.

Interestingly, the degree of the Gd^{III}-dependent stabilization of duplex **7·8** containing one **dcaU–dcaU** pair (ΔT_m = + 16.1 °C) was comparable with that of a duplex with three **U^{OH}–U^{OH}** pairs (ΔT_m = + 18.3 °C) previously reported [1]. This result reflects the improved coordination ability of the **dcaU** base compared to **U^{OH}**. While the **U^{OH}** base is a bidentate ligand with an *O*,*O*-coordination mode, the **dcaU** base is expected to behave as a tetradentate ligand through coordination of the IDA ligand and 4-carbonyl oxygen. This feature of the **dcaU** base allowed for the formation of a single **dcaU**–Gd^{III}–**dcaU** pair within duplexes, which is stable enough to compensate for structural distortion caused by metal complexation.

Thus, duplex melting analysis revealed that duplex **7·8** with a **dcaU–dcaU** pair was thermally stabilized by the formation of a metal-mediated **dcaU**–Gd^{III}–**dcaU** base pair (ΔT_m = + 16.1 °C). This significant stabilization of DNA duplexes by the single **dcaU**–Gd^{III}–**dcaU** pairing would be useful for metal-dependent control of DNA assemblies.

3.5 Base-Pairing Behaviors of *N*,*N*-Dicarboxymethyl-5-Aminouracil Bases with Natural Nucleobases

Since the **dcaU** base retains the hydrogen-bonding site of natural thymine or uracil bases, it was expected to form a hydrogen-bonded base pair with a natural adenine (A). Herein the base-pairing property of **dcaU** with natural nucleobases was examined by duplex melting analysis.

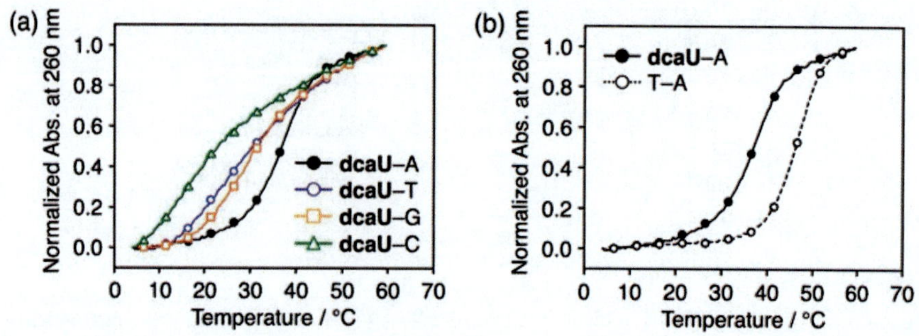

Fig. 3.26 **a** Melting curves of DNA duplexes containing a **dcaU**–N pair (N = A, T, G, or C). **b** Melting curves of duplexes containing a **dcaU**–A (**7·8A**) or T–A pair (**7T·8A**). [duplex] = 2.0 μM in 10 mM HEPES–NaOH buffer (pH 7.0), 100 mM NaCl, 0.2 °C/min

First, thermal stability of duplex **7·8N** containing a **dcaU**–N (N = A, T, G, or C) pair was evaluated. As shown in Fig. 3.26a, duplex **7·8A** with a **dcaU**–A pair exhibited the highest thermal stability among the four duplexes. From this result, it was suggested that the **dcaU** base forms a hydrogen-bonded **dcaU**–A pair in a manner similar to a natural T–A pair. The selective base pairing of **dcaU** with natural A bases was consistent with the previous studies using other 5-modified uracil bases such as $\mathbf{U^{OH}}$ and **caU** [1–4]. It should be noted that the melting temperature of duplex **7·8A** (T_m = 38.0 °C) was lower than that of duplex **7T·8A** containing a T–A pair (47.9 °C) (Fig. 3.26b). This lower stability can result from the electrostatic repulsion between the negatively charged IDA ligand and phosphate backbones.

Fig. 3.27a describes the effects of Gd^{III} ions on the stability of duplex **7·8A**. In the presence of 1 equiv. of Gd^{III} ions, the duplex was slightly destabilized (T_m = 34.5 °C, ΔT_m = –3.5 °C). However, duplex **7T·8A** containing a natural T–A pair was not affected by the addition of Gd^{III} ions at all (Fig. 3.27b). These results suggested that the hydrogen bonding between **dcaU** and A bases was weakened by 1:1 complexation of the **dcaU** base and a Gd^{III} ion (Fig. 3.27c). It is also worth mentioning that the Gd^{III}-dependent thermal destabilization was further enhanced by the addition of excess Gd^{III} ions. This result suggests that the coordination strength of the 1:1 complexation was not so high as that of the 2:1 complexation. This Gd^{III}-dependent destabilization was also observed in the previous study using a duplex containing $\mathbf{U^{OH}}$–A pairs [1].

Melting temperatures of duplexes containing the **dcaU** base were summarized in Table 3.3. In the absence of Gd^{III} ions, duplex **7·8A** with a **dcaU**–A pair (T_m = 38.0 °C) showed much higher stability than duplex **7·8** containing a **dcaU**–**dcaU** pair (22.5 °C). On the other hand, in the presence of equimolar Gd^{III} ions, duplex **7·8** became more stable (38.6 °C) than duplex **7·8A** (34.5 °C) due to the formation of the **dcaU**–Gd^{III}–**dcaU** pair and the destabilization of the **dcaU**–A pair. Thus, the relative thermal stability of the two duplexes was inverted by the addition of Gd^{III} ions, indicating the switchable base-pairing behaviors of **dcaU**. This Gd^{III}-responsive base pair switching between the **dcaU**–A and **dcaU**–Gd^{III}–**dcaU**

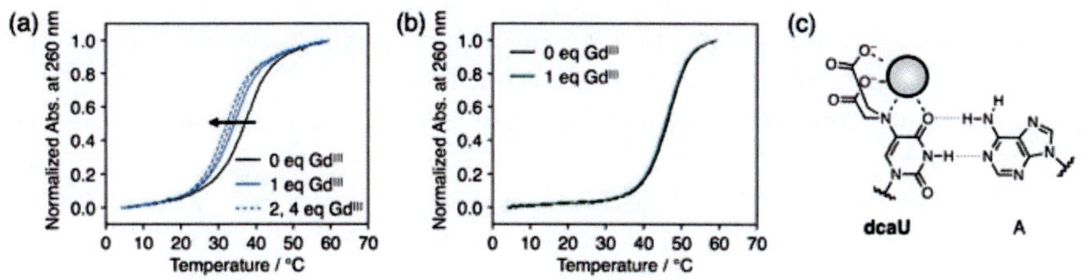

Fig. 3.27 **a**, **b** Melting curves of DNA duplexes containing a **a** **dcaU**–A (**7**·**8A**) or **b** T–A pair (**7T**·**8A**) in the middle in the absence and presence of Gd^{III} ions. [duplex] = 2.0 μM, $[Gd^{III}]$/[duplex] = 0 (black solid line), 1 (blue solid line), 2, and 4 (blue dotted lines) in 10 mM HEPES–NaOH buffer (pH 7.0), 100 mM NaCl, 0.2 °C/min. **c** Proposed structure of the **dcaU**–A pair in the presence of Gd^{III} ions

Table 3.3 Melting temperatures (T_m) of **dcaU**-containing duplexes in the absence and presence of Gd^{III} ions[a]

Gd^{III}/eq	Duplex **7**·**8** (**dcaU**–**dcaU**)		Duplex **7**·**8A** (**dcaU**–A)	
	T_m/°C	ΔT_m/°C	T_m/°C	ΔT_m/°C
0	22.5 ± 0.6 °C	–	38.0 ± 0.4 °C	–
1	38.6 ± 0.6 °C	+ 16.1 ± 0.4 °C	34.5 ± 0.6 °C	− 3.5 ± 1.1 °C

[a]Averaged values calculated from three independent measurements are shown.
Error values stand for standard errors.

pairs was expected to be applicable to Gd^{III}-dependent control of DNA hybridization preferences.

3.6 Metal-Dependent Switching of the DNA Hybridization Partner through Base Pair Switching of *N*,*N*-Dicarboxymethyl-5-Aminouracil Bases

As described in Sects. 3.4 and 3.5, thermal stability of duplex **7**·**8** containing a **dcaU**–**dcaU** pair and duplex **7**·**8A** containing a **dcaU**–A pair was inverted by the addition of Gd^{III} ions. This result suggests that the base-pairing behavior of **dcaU** can be reversibly switched between **dcaU**–A and **dcaU**–Gd^{III}–**dcaU**. The Gd^{III}-dependent base pair switching of the **dcaU** base was further applied to control the hybridization preference of the **dcaU**-containing DNA strand.

First, the hybridization preference of strand **7** with a **dcaU** base was investigated in the presence of two complementary strands **8** and **8A′** containing **dcaU** and A bases, respectively, at the opposite site of **dcaU** (Fig. 3.28a). Strand **8A′** was designed to be 2-base shorter than the others to enhance the efficiency of the strand exchange. The equimolar mixture of the three strands were annealed in the absence and presence of Gd^{III} ions and the hybridization products were analyzed by native polyacrylamide gel electrophoresis (PAGE). To quantitatively evaluate the hybridization behaviors, the 5′ terminus of strand **8A′** was labeled by a FAM fluorophore. The result showed that duplex **7**·**8A′** was formed in 84% yield in the absence of Gd^{III} ions. In the presence of 1 equiv. of Gd^{III} ions, 55% of strand **8A′** was released, suggesting the formation of duplex **7**·**8** containing the **dcaU**–Gd^{III}–**dcaU** pair (Fig. 3.28b). The band of duplex **7**·**8** was also observed in the presence of Gd^{III} ions after staining the gel by SYBR Gold (Fig. 3.28c). This result indicates that the hybridization partner of **dcaU**-containing strand **7** was exchanged from **8A′** to **8**. The Gd^{III} titration experiment revealed that the amount of the released strand **8A′** was maximal when 1 equiv. of Gd^{III} ions was added (Figs. 3.28 and 3.29). This stoichiometric response was observed because the addition of excess Gd^{III} ions resulted in destabilization of duplex **7**·**8**, which was observed in the duplex melting analysis (Fig. 3.23).

With a view to controlling the hybridization behavior in a reversible manner, I have additionally examined whether the removal of Gd^{III} ions by chelating reagents

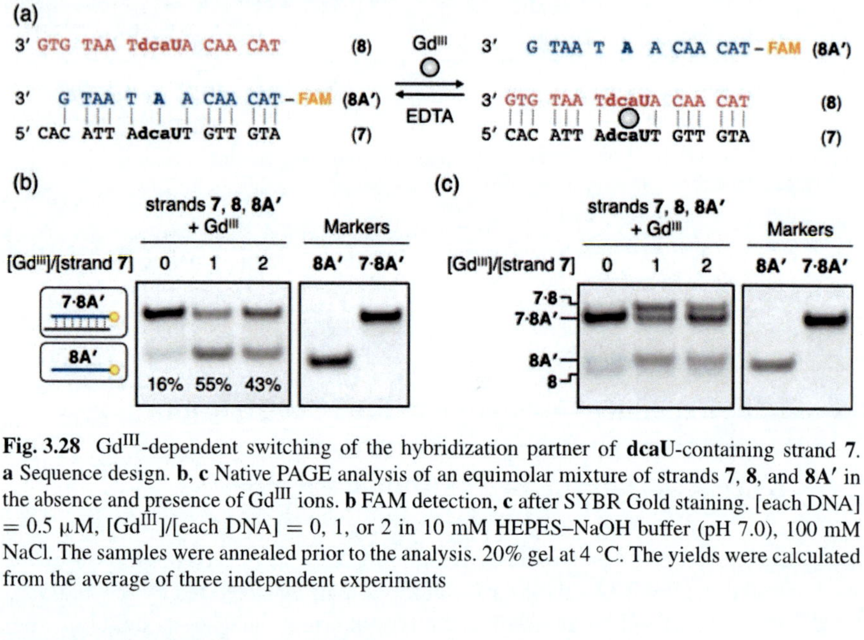

Fig. 3.28 Gd^{III}-dependent switching of the hybridization partner of **dcaU**-containing strand **7**. **a** Sequence design. **b**, **c** Native PAGE analysis of an equimolar mixture of strands **7**, **8**, and **8A′** in the absence and presence of Gd^{III} ions. **b** FAM detection, **c** after SYBR Gold staining. [each DNA] = 0.5 μM, [Gd^{III}]/[each DNA] = 0, 1, or 2 in 10 mM HEPES–NaOH buffer (pH 7.0), 100 mM NaCl. The samples were annealed prior to the analysis. 20% gel at 4 °C. The yields were calculated from the average of three independent experiments

Fig. 3.29 Yields of duplex **7**·**8A′** and single strand **8A′** in the presence of varying amounts of Gd^{III} ions. The yields were calculated from the average of three independent experiments. Error bars stand for standard errors

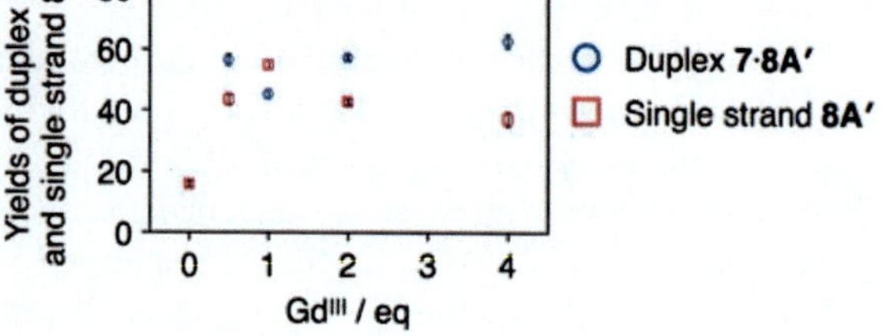

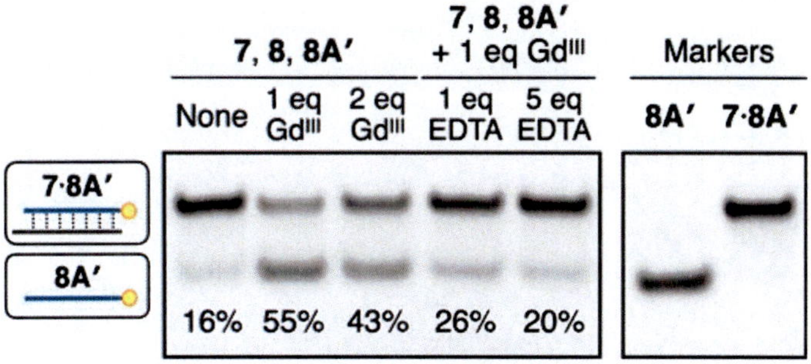

Fig. 3.30 Gd^{III}-dependent reversible switching of the hybridization partner of **dcaU**-containing strand **7** in response to Gd^{III} ions and EDTA. [each DNA] = 0.5 μM, [Gd^{III}]/[each DNA] = 0, 1, or 2, [EDTA]/[each DNA] = 0, 1, or 5 in 10 mM HEPES–NaOH buffer (pH 7.0), 100 mM NaCl. The samples were annealed prior to the analysis. 20% gel at 4 °C

restores the Gd^{III}-free state, which mainly contains duplex **7**·**8A**′ (84%). As expected, duplex **7**·**8A**′ was reproduced in 74% yield by the addition of 1 equiv. of EDTA to the Gd^{III}-containing sample (Fig. 3.30). Furthermore, duplex **7**·**8A**′ was more efficiently restored (80% yield) in the presence of 5 equiv. of EDTA. The excess amount of EDTA was needed possibly because the coordination ability of the **dcaU** base toward Gd^{III} ions is so high as that of EDTA.

The hybridization preference of **dcaU**-containing strand **7** was found to be controlled even under isothermal conditions (Fig. 3.31a). To evaluate the Gd^{III}-responsive exchange of the hybridization partner of strand **7**, an adenine base of strand **8A**′ was replaced by 2-aminopurine (**AP**), a fluorescent nucleobase analog [18]. The **AP** base forms a hydrogen-bonded **AP**–T base pair with a natural thymine base to show quenched emission (Fig. 3.31b). Therefore, the release of single strand **8A**′ can be detected by the emission increase of the **AP** base. After annealing the equimolar mixture of three strands **7**, **8**, and **8A**′, 1 equiv. of Gd^{III} ions was added and the fluorescence intensity was recorded in constant intervals (Fig. 3.31c, red line). The emission intensity was immediately increased by the addition of Gd^{III} ions and reached plateau within 15 min. This result indicated the release of strand **8A**′ and the formation of duplex **7**·**8** containing the **dcaU**–Gd^{III}–**dcaU** pair. When strands **7T**, **8T**, and **8A**′ were used by replacing the **dcaU** bases with natural T bases, no fluorescence change was observed upon addition of Gd^{III} ions (Fig. 3.31c, black line). These results showed that the Gd^{III}-responsive strand exchange was caused by the **dcaU**–Gd^{III}–**dcaU** complexation.

Then the rate of the Gd^{III}-responsive strand exchange using **dcaU**-containing strands (Fig. 3.32a) was compared with that of standard DNA strand exchange using natural strands (Fig. 3.32b). Strand **8A**, which is perfectly complementary to strand **7T**, was added to the preannealed duplex **7T**·**8A**′. The **AP** fluorescence was gradually increased, showing the release of strand **8A**′ and the formation of the other duplex **7T**·**8A** (Fig. 3.32c). However, the rate of the strand exchange was much slower than the Gd^{III}-responsive change using the **dcaU**-containing strands.

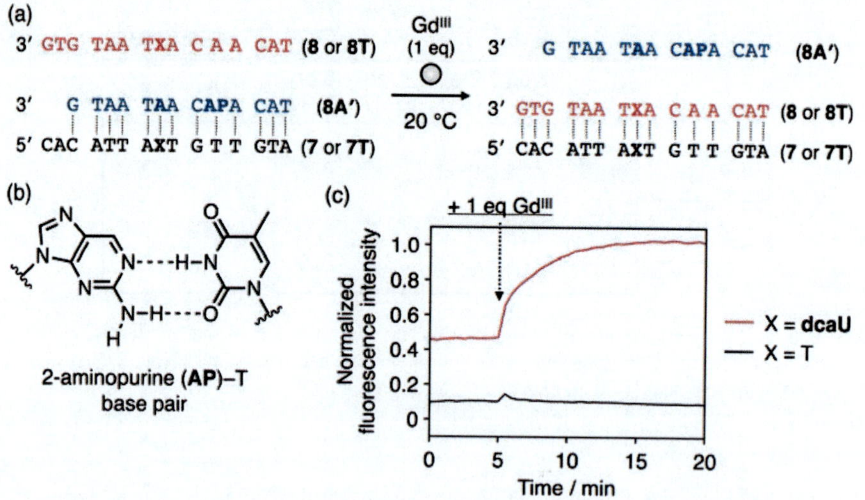

Fig. 3.31 Gd^{III}-responsive switching of the hybridization partner of **dcaU**-containing strand **7** under isothermal conditions. **a** Sequence design. **b** Chemical structure of the 2-aminopurine (**AP**)–T base pair. **c** Time-course fluorescence analysis of the Gd^{III}-responsive strand exchange. [each DNA] = 0.5 μM, [Gd^{III}]/[each DNA] = 1 in 10 mM HEPES–NaOH buffer (pH 7.0), 100 mM NaCl. After annealing the mixture of the three strands, Gd^{III} (1 equiv.) was added. $\lambda_{ex} = 303$ nm, $\lambda_{em} = 371$ nm, 20 °C. Observed fluorescence intensity was normalized by that of duplex **7**·**8A′** or **7T**·**8A′** with the **AP**–T pair and **AP**-containing single strand **8A′**

This result suggests that the Gd^{III}-dependent strand exchange was not driven by branch migration observed in the conventional strand displacement reactions, but by (i) Gd^{III}-triggered destabilization of duplex **7**·**8A′** and (ii) the subsequent dissociation of duplex **7**·**8A′** and formation of duplex **7**·**8**.

Reversible exchange of the hybridization partner of strand **7** was also demonstrated by adding EDTA to scavenge Gd^{III} ions (Fig. 3.33a). Twenty minutes after the addition of Gd^{III} ions, 5 equiv. of EDTA were added to completely remove Gd^{III} ions from duplex **7**·**8**. As a result, the fluorescence intensity was gradually decreased to the level similar to the Gd^{III}-free condition, indicating the regeneration of duplex **7**·**8A′** (Fig. 3.33b). These results showed that the hybridization preference of **dcaU**-containing strand **7** was reversibly switched by the addition of Gd^{III} ions and EDTA.

Furthermore, the repeated exchange of the hybridization partner was demonstrated by the sequential addition of Gd^{III} ions and EDTA (Fig. 3.33c). To a preannealed mixture of strands **7**, **8**, and **8A′**, Gd^{III} ions and EDTA were alternately added in certain time intervals. The fluorescence change was repeatedly switched, showing the reversible exchange of the hybridization partner of strand **7**, although the efficiency of the strand exchange gradually declined. Hence, the hybridization partner of **dcaU**-containing strand **7** was reversibly converted in response to Gd^{III} ions, even under isothermal conditions.

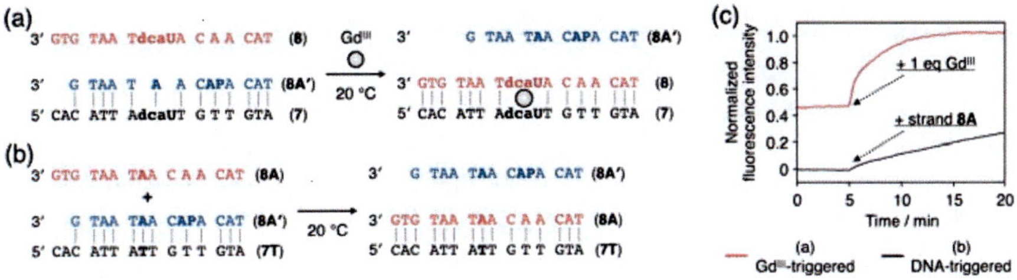

Fig. 3.32 Comparison of the speed of the Gd^{III}-responsive strand exchange with that of a toehold-mediated strand exchange using natural DNA strands. **a** Scheme of the Gd^{III}-responsive strand exchange using **dcaU**-containing strands. After annealing the mixture of the three strands, Gd^{III} (1 equiv.) was added. **b** Scheme for toehold-mediated strand exchange using natural DNA strands. After annealing the mixture of strands **7T** and **8A′**, strand **8A** was added. **c** Time-course fluorescence analysis of the strand exchange. [each DNA] = 0.5 μM, [Gd^{III}]/[each DNA] = 0 or 1 in 10 mM HEPES–NaOH buffer (pH 7.0), 100 mM NaCl. λ_{ex} = 303 nm, λ_{em} = 371 nm, 20 °C. Observed fluorescence intensity was normalized by that of duplex **7**·**8A′** or **7T**·**8A′** with the **AP**–T pair and **AP**-containing single strand **8A′**

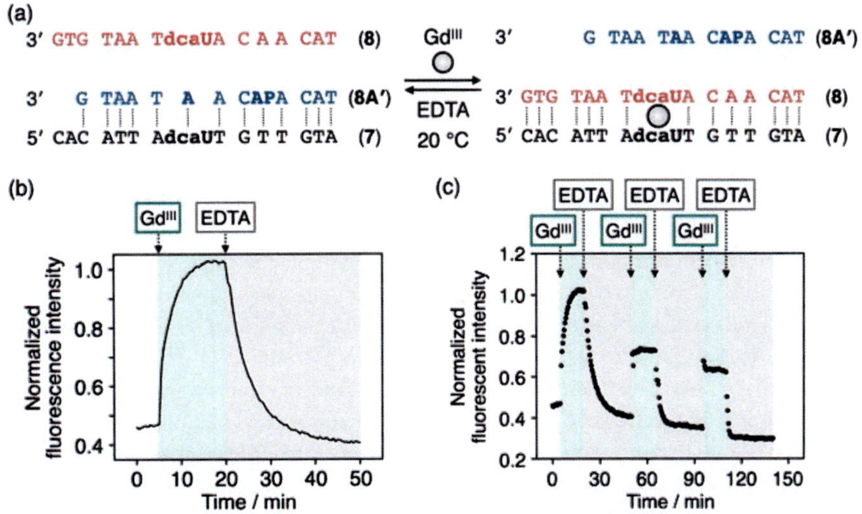

Fig. 3.33 Reversible switching of the hybridization partner of **dcaU**-containing strand **7** by the addition and removal of Gd^{III} ions. **a** Sequence design. **b** Reversible strand exchange in response to Gd^{III} and EDTA. After annealing the mixture of the three strands, Gd^{III} ions (1 equiv.) and EDTA (5 equiv.) were sequentially added. **c** Repeated exchange of the hybridization partner of strand **7** by the sequential addition of Gd^{III} and EDTA. To a preannealed mixture of the three strands, Gd^{III} (1 equiv.), EDTA (5 equiv.), Gd^{III} (5 equiv.), EDTA (25 equiv.), Gd^{III} (25 equiv.), and EDTA (125 equiv.) were alternately added. [each DNA] = 0.5 μM in 10 mM HEPES–NaOH buffer (pH 7.0), 100 mM NaCl. λ_{ex} = 303 nm, λ_{em} = 371 nm, 20 °C. Observed fluorescence intensity was normalized by that of duplex **7**·**8A′** with the **AP**–T pair and **AP**-containing single strand **8A′**

In summary, the hybridization preference of strand **7** was altered through the base pair switching of **dcaU** between the **dcaU**–A and **dcaU**–Gd^{III}–**dcaU** pairs. By the addition and removal of Gd^{III} ions, the hybridization partner of the **dcaU**-containing strand was reversibly switched under isothermal conditions. The base pair switching of **dcaU** is expected to be further applicable to metal-responsive control of higher-ordered DNA assemblies and DNA nanodevices.

3.7 Discussion

I have developed a novel *N*,*N*-dicarboxymethyl-5-aminouracil (**dcaU**) nucleobase possessing an IDA ligand at the 5 position as a bifacial nucleobase that forms both a hydrogen-bonded **dcaU**–A base pair and a metal-mediated **dcaU**–Gd^{III}–**dcaU** pair. The **dcaU** nucleoside was synthesized from 5-bromo-2′-deoxyuridine in good yields and successfully incorporated into DNA strands by standard phosphoramidite chemistry. UV–vis titration experiments and ESI mass spectrometry confirmed 2:1 complexation of the **dcaU** nucleosides and a Gd^{III} ion. Then the base-pairing behavior of the **dcaU** base within DNA duplexes was examined by duplex melting analysis. Duplex **7**·**8** with a **dcaU**–**dcaU** pair in the middle was thermally stabilized by the addition of Gd^{III} ions, indicating the formation of a single metal-mediated **dcaU**–Gd^{III}–**dcaU** base pair. Base-pairing behaviors of **dcaU** with natural nucleobases were subsequently studied. Among four duplexes containing a **dcaU**–N pair (N = A, T, G, or C), duplex **7**·**8A** with a **dcaU**–A pair showed the highest stability, suggesting that **dcaU** can form a hydrogen-bonded **dcaU**–A pair within the duplex. Duplex **7**·**8A** was found to be destabilized by the addition of Gd^{III} ions, possibly because of weakened hydrogen bonding between **dcaU** and A. Finally, the bifacial base pairing of the **dcaU** base was applied to control the hybridization preference of **dcaU**-containing strand **7** (Fig. 3.34). The addition of Gd^{III} ions altered the hybridization partner of strand **7**, converting duplex **7**·**8A′** containing the **dcaU**–A pair to duplex **7**·**8** with the **dcaU**–Gd^{III}–**dcaU** pair. It was also possible to regenerate duplex **7**·**8A′** by removing Gd^{III} ions with EDTA. Furthermore, it was demonstrated that the hybridization preference of strand **7** can be repeatedly switched for several cycles by the addition and removal of Gd^{III} ions under isothermal conditions.

In the previous study using $\mathbf{U^{OH}}$ bases, it was difficult to construct a single $\mathbf{U^{OH}}$–Gd^{III}–$\mathbf{U^{OH}}$ base pair within DNA duplexes because the coordination strength of $\mathbf{U^{OH}}$

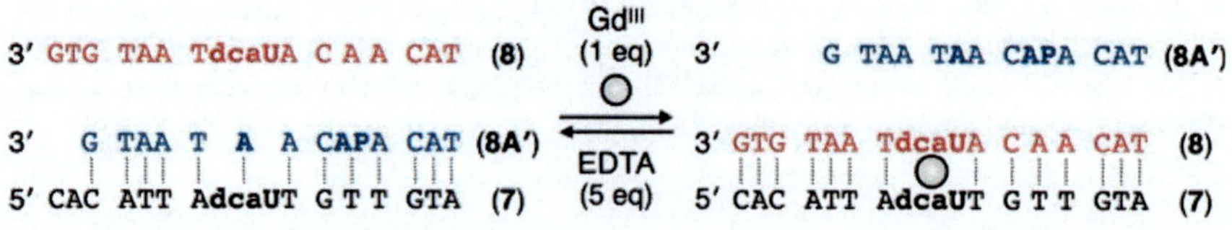

Fig. 3.34 Gd^{III}-responsive strand exchange driven by the base pair switching of the **dcaU** base

was not enough to overcome the severe structural distortion caused by metal complexation. In this study, the tridentate IDA ligand was introduced instead of the hydroxy group to design the **dcaU** base. As a result, the **dcaU** base functioned as a tetradentate ligand and the single metal-mediated **dcaU**–Gd^{III}–**dcaU** pair was stably formed inside DNA duplexes. It can be because the tetradentate nature of **dcaU** was suitable for coordination to Gd^{III} ions and the binding strength was improved compared with the bidentate **U^{OH}** base. In fact, thermal stabilization of duplexes by the single **dcaU**–Gd^{III}–**dcaU** pairing ($\Delta T_m = + 16.1$ °C) was almost comparable with that by the formation of three consecutive **U^{OH}**–Gd^{III}–**U^{OH}** pairs ($\Delta T_m = + 18.3$ °C) [1]. Accordingly, this study demonstrated that the binding strength and metal selectivity of the 5-modified uracil bases can be tuned by carefully designing ligands for modification.

Most of the reported metal-mediated base pairs have flat structures to be fit within the base-stacking environment of DNA duplexes [19–21]. In contrast, the formation of the **dcaU**–Gd^{III}–**dcaU** complex was allowed within DNA duplexes in spite of its bulky and nonplanar structure. Considering the fact that modifications at the 5 position of pyrimidine bases tend to project into the major groove of DNA duplexes [22], the bulky **dcaU**–Gd^{III}–**dcaU** complex may be stably formed in the duplex groove. It should be noted that the ΔT_m value of the duplex containing the **dcaU**–Gd^{III}–**dcaU** pair (+ 16.1 °C) was also in the level similar to a duplex with a **H**–Cu^{II}–**H** base pair (+ 13.1 °C), which is one of the most studied metal-mediated base pairs with various applications [23–26]. Therefore, the **dcaU**–Gd^{III}–**dcaU** pairing would be applicable to metal-responsive control of DNA assemblies and functional nucleic acids. In addition, metal-mediated base pairs have been usually designed to take linear and square-planar geometry with monodentate and bidentate ligands, respectively. While many stable metal-mediated base pairs have been developed based on this strategy, the variety of bridging metal ions have been limited. Herein I designed **dcaU** as a tetradentate ligand and demonstrated its selective complexation with Gd^{III} ions. Taken all together, this study will greatly expand the chemical diversity of metal-mediated base pairs.

The base pair switching of 5-modified uracil bases has been expected to be applicable to metal-responsive control of DNA assemblies. In fact, Gd^{III}-responsive DNA strand displacement reactions (SDRs) were successfully developed using the base pair switching of **U^{OH}**, as demonstrated in Chap. 2. However, the sequence flexibility has been limited because previously developed **U^{OH}** and **caU** bases had to be consecutively incorporated into DNA strands for the metal-mediated base pair formation. In contrast, the **dcaU** base forms a stable **dcaU**–Gd^{III}–**dcaU** pair within DNA duplexes, enabling the Gd^{III}-dependent base pair switching by incorporating just a single **dcaU** base. Therefore, it is possible to endow natural DNA assemblies with metal responsiveness by replacing any T base with **dcaU**. Thus, this study will enlarge the applicability of the metal-responsive base pair switching using 5-modified uracil bases to construct dynamic DNA systems.

3.8 Experimental Section

General

All organic syntheses were conducted under an argon atmosphere using commercial dehydrated solvents (Wako Pure Chemical Industries). The chemical reagents were purchased from Wako Pure Chemical Industries, Tokyo Chemical Industry (TCI), and Sigma Aldrich, and used without further purification. Merck Silica Gel 60 (80–230 or 230–400 mesh) or amine-modified silica gel NH-DM1020 (Fuji Silysia) was used for silica gel column chromatography. NMR spectroscopy was conducted using a Bruker AVANCE 500 spectrometer (500 MHz for ^{1}H, 126 MHz for ^{13}C, and 202 MHz for ^{31}P). The obtained NMR spectra were referenced to the signal of tetramethylsilane (TMS) in $CDCl_3$ (δ 0 ppm) or to the residual solvent signals in CD_3OD (δ 3.31 ppm for ^{1}H, 49.00 ppm for ^{13}C). Electrospray ionization-time-of-flight (ESI-TOF) mass spectra were measured by a Waters LCT Premier XE or a Bruker Compact System. DNA strands without **dcaU** bases were purchased from Japan Bio Services at HPLC purification grade. Metal sources were purchased from FUJIFILM Wako Pure Chemical Industries ($FeCl_3 \cdot 6H_2O$ (99.9% purity), $NiCl_2 \cdot 6H_2O$ (99.9%), and $CuCl_2 \cdot 2H_2O$ (99.9%)) and Soekawa Chemical Co. ($ZnSO_4 \cdot 7H_2O$ (99.9%) and $GdCl_3 \cdot 6H_2O$ (99.9%)).

Synthesis of Compound 3

5-Bromo-2′-deoxyuridine (768 mg, 2.50 mmol) was dissolved in benzylamine (5.5 mL, 50.0 mmol, 20 eq) and the solution was stirred at 90 °C for 3 h. The reaction mixture was concentrated under reduced pressure. Excess benzylamine in the solution was removed by co-evaporation with toluene. The residue was dissolved in water (100 mL) and washed with CH_2Cl_2 (100 mL). The organic layer was extracted by water (100 mL) and the aqueous layer was collected. 28% NH_3 aq (25 mL) was added to the combined aqueous layer and the solution was washed again with CH_2Cl_2 (50 mL × 4). After concentrated under reduced pressure, the crude product of 5-benzylamino-2′-deoxyuridine (859 mg) was obtained. The residue was dissolved in methanol (38 mL) and mixed with 10% Pd/C (266 mg, 0.25 mmol). The obtained suspension was vigorously stirred under a H_2 atmosphere at room temperature for 2 h. The Pd/C powder was removed by filtration, followed by the concentration under reduced pressure to afford 5-amino-2′-deoxyuridine (672 mg, containing impurities). A part of the obtained crude 5-amino-2′-deoxyuridine (120 mg, containing impurities) was dissolved in methanol (1.9 mL) followed by the addition of methyl bromoacetate (93.6 μL, 0.989 mmol, 2.0 eq) and 2,6-lutidine (115 μL, 0.989 mmol, 2.0 eq). After the reaction mixture was stirred at 60 °C for 2 h, methyl bromoacetate (328 μL, 3.46 mmol, 7.0 eq in total) was added again in four portions for over 25 h. The reaction mixture was further stirred at 60 °C for 17 h. After removing the solvent under reduced pressure, the residue was purified by silica gel column chromatography ($CHCl_3$:CH_3OH = 50:1–1:1) and the target compound **3** was obtained as a colorless solid (65.2 mg, 0.168 mmol, 38% in two steps).

^{1}H NMR (500 MHz, CD_3OD, 300 K): δ 7.59 (s, 1H), 6.31 (dd, $J = 6.7, 6.7$ Hz, 1H), 4.41–4.38 (m, 1H), 4.06–3.98 (m, 4H), 3.93–3.91 (m, 1H), 3.79–3.71 (m, 2H), 3.70 (s, 6H), 2.28–2.16 (m, 2H).

^{13}C NMR (126 MHz, CD_3OD, 300 K): δ 172.82, 162.90, 151.15, 128.43, 126.09, 89.01, 86.43, 72.41, 62.97, 54.23, 52.28, 41.26.

HRMS (ESI-TOF) *m/z*: $[M + Na]^+$ calcd for $C_{15}H_{21}N_3NaO_9$ 410.1170, found 410.1189.

Synthesis of Compound 4·Na$^+_2$

NaOH aqueous solution (0.1 M, 674 μL, 0.067 mmol, 2.0 eq) was added to nucleoside **3** (13 mg, 0.034 mmol). The obtained solution was incubated at 60 °C for 22 h and concentrated under reduced pressure. The disodium salt of the target compound **4** was obtained as a colorless solid (14 mg, 0.035 mmol, quant.).

^{1}H NMR (500 MHz, CD_3OD, 300 K): δ 7.75 (s, 1H), 6.28 (dd, $J = 6.5, 6.5$ Hz, 1H), 4.42–4.39 (m, 1H), 3.90–3.82 (m, 2H), 3.75–3.72 (m, 1H), 3.50–3.42 (m, 4H), 2.25–2.23 (m, 2H).

^{13}C NMR (126 MHz, CD_3OD, 300 K): 177.91, 165.26, 151.47, 130.45, 127.62, 88.93, 86.54, 71.99, 62.67, 58.99, 41.38.

HRMS (ESI-TOF) *m/z*: $[M + H + 2Na]^+$ calcd for $C_{13}H_{16}N_3Na_2O_9$ 404.0676, found 404.0678.

Synthesis of Compound 5

Nucleoside **3** (305 mg, 0.676 mmol) and 4-dimethylaminopyridine (8.3 mg, 0.068 mmol, 0.10 eq) were dissolved in pyridine (3.4 mL). After mixed with a solution of 4,4′-dimethoxytrityl chloride (412 mg, 1.22 mmol, 1.8 eq) in pyridine (3.4 mL), the reaction solution was stirred at room temperature for 2 h. After the addition of methanol (1.6 mL) to quench the reaction, the solution was concentrated under reduced pressure. The obtained residue was purified by silica gel column chromatography ($CHCl_3$:CH_3OH = 100:1–10:1) and the target compound **5** was obtained as a colorless foam (274 mg, 0.398 mmol, 59%).

^{1}H NMR (500 MHz, CD_3OD, 300 K): δ 7.47–7.45 (m, 2H), 7.36–7.33 (m, 4H), 7.29–7.26 (m, 2H), 7.21–7.18 (m, 2H), 6.90–6.80 (m, 4H), 6.26 (dd, $J = 6.6, 6.6$ Hz, 1H), 4.33–4.31 (m, 1H), 4.05–4.02 (m, 1H), 3.90–3.79 (m, 4H), 3.76 (s, 6H), 3.53 (s, 6H), 3.38–3.28 (m, 2H), 2.32–2.27 (m, 1H), 2.15–2.10 (m, 1H).

^{13}C NMR (126 MHz, CD_3OD, 300 K): δ 172.40, 162.97, 160.16, 151.10, 146.31, 137.16, 137.10, 131.36, 129.85, 129.38, 128.85, 127.86, 125.95, 114.18, 114.17, 87.72, 87.29, 86.38, 72.67, 65.31, 55.71, 54.41, 52.19, 40.51.

HRMS (ESI-TOF) *m/z*: $[M + Na]^+$ calcd for $C_{36}H_{39}N_3NaO_{11}$ 712.2477, found 712.2455.

Synthesis of Compound 6

A solution of nucleoside **5** (146 mg, 0.212 mmol) in CH_2Cl_2 (1.3 mL) was mixed with 2-cyanoethyl-*N,N,N',N'*-tetraisopropylphosphorodiamidite (162 μL, 0.510 mmol, 2.4 eq), diisopropylamine (112 μL, 0.786 mmol, 3.7 eq), and tetrazole (55.1 mg, 0.786 mmol, 3.7 eq). The reaction was performed at room temperature for 75 min and quenched with benzyl alcohol polymer bound (321 mg, 0.321–0.642 mmol, 1.5–3.0 eq). The polymer was removed by filtration, and the reaction mixture was concentrated under reduced pressure. The obtained residue was purified by silica gel column chromatography ($CHCl_3$:CH_3OH = 100:0–10:1) to afford the target compound **6** as a colorless foam (122 mg, 0.138 mmol, 65%, diastereomeric mixture). After characterized by ^{1}H and ^{31}P NMR spectroscopy and mass spectrometry, phosphoramidite **6** was immediately used for solid-phase DNA synthesis by an automated synthesizer.

^{1}H NMR (500 MHz, $CDCl_3$, 300 K): δ 7.98 (br, 1H), 7.47–7.43 (m, 2H), 7.34–7.18 (m, 8H), 6.85–6.80 (m, 4H), 6.27–6.23 (m, 1H), 4.54–4.48 (m, 1H), 4.16–4.12 (m, 1H), 3.88–3.69 (m, 11H), 3.66–3.58 (m, 3H), 3.55 (s, 6H), 3.40–3.32 (m, 2H), 2.61 (t, 1H, J = 6.4 Hz), 2.51–2.47 (m, 0.5H), 2.44–2.39 (m, 1.5H), 1.18–1.15 (m, 9.3H), 1.07 (d, 2.7H, J = 6.8 Hz).

^{31}P NMR (202 MHz, $CDCl_3$, 300 K): 149.23, 148.80 (diastereomers).

HRMS (ESI-TOF) m/z: $[M + Na]^+$ calcd for $C_{45}H_{56}N_5NaO_{12}P$ 912.3555, found 912.3588.

DNA Synthesis

DNA strands containing *N*,*N*-dicarboxymethyl-5-aminouracil (**dcaU**) nucleosides were synthesized on an Applied Biosystems 394 DNA synthesizer on a 1-μmol scale in a DMTr-on mode using ultramild deprotection phosphoramidites and reagents (Glen Research). DNA synthesis was conducted following the standard procedure except for an extended coupling time for the **dcaU** nucleoside (15 min). The synthesized strands were cleaved from the solid support by treating them with a 0.3 M NaOH aqueous solution (1.0 mL) at room temperature for 2 h. After the addition of a 0.3 M NaOH aqueous solution (1.0 mL), the supernatant was taken out and incubated at 37 °C for 22 h to remove protecting groups on the nucleobases and the phosphate backbones. The obtained DNA strands were first detritylated and purified by a Glen-Pak cartridge (Glen Research). After further purification by reverse-phase HPLC (Waters XBridge Oligonucleotide OST C18 column, buffers: A = MeCN, B = 0.1 M TEAA (pH 7.0) + 2% MeCN, 60 °C), all DNA strands were identified by ESI-TOF mass spectrometry. The amount of the DNA strands was calculated based on the UV absorbance value at 260 nm. The molar extinction coefficients (ε_{260}) of the strands were determined based on the nearest-neighbor method ($\varepsilon_{260} = 5.12 \times 10^3$ M^{-1} cm^{-1} and 1.00×10^3 M^{-1} cm^{-1} for the **dcaU** and AP nucleosides, respectively). The FAM-labeled strand was quantified based on UV absorbance at 495 nm ($\varepsilon_{495} = 7.50 \times 10^3$ M^{-1} cm^{-1} for FAM).

Strand **7**. 5′ - CAC ATT A**dcaU**T GTT GTA - 3′. HPLC retention time: 32.1 min (gradient: 4% A (0 min), 6.5% A (40 min)). ESI MS: m/z calcd for $[C_{151}H_{191}N_{51}O_{95}P_{14} - 4H]^{4-}$: 1167.51, found: 1167.45.

Strand **8**. 5′ - TAC AAC A**dcaU**T AAT GTG - 3′. HPLC retention time: 31.6 min (gradient: 4% A (0 min), 6.5% A (40 min)). ESI MS: m/z calcd for $[C_{151}H_{189}N_{57}O_{91}P_{14} - 4H]^{4-}$: 1171.70, found: 1171.96.

Determination of the Molar Extinction Coefficient of the dcaU Nucleoside

Fully deprotected nucleoside **4** was dissolved in 10 mM HEPES–NaOH buffer (pH 7.0) and a 100-μM sample solution was prepared. UV–vis absorption spectrum was recorded on a V-730 spectrophotometer (JASCO) at 25 °C with a path length of 1 cm. Based on the recorded absorbance value at 260 nm, the molar extinction coefficient (ε_{260}) of **dcaU** nucleoside was calculated to be 5.12×10^3 M^{-1} cm^{-1}.

Metal Complexation Study of the dcaU Nucleoside

Nucleoside **4** was mixed with varying concentrations of Gd^{III} ions in 10 mM HEPES–NaOH buffer (pH 7.0) to prepare the sample solutions for UV–vis absorption measurement. The solutions were incubated at 25 °C for at least 1 h prior to the measurement. UV absorption spectra were measured by a V-730 spectrophotometer (JASCO) at 25 °C with a path length of 0.3 cm.

ESI-TOF mass spectra were recorded on a Waters LCT Premier XE. The sample solution was prepared by mixing nucleoside **4** (100 μM) with 0.5 equiv. of Gd^{III} ions in 10 mM NH_4OAc buffer (pH 7.0). The measurement was performed in the positive mode.

Duplex Melting Analysis

DNA strands (2.0 μM each) were mixed with metal ions in 10 mM HEPES–NaOH buffer (pH 7.0) containing 100 mM NaCl. The prepared samples were annealed (from 60 °C to 4 °C, − 1.0 °C/min) prior to the measurements. Absorbance at 260 nm (A_{260}) was recorded on UV-1800 and UV-1900 spectrophotometers (Shimadzu) equipped with a TMSPC-8 temperature controller. The temperature was raised from 4 °C to 60 °C at the rate of 0.2 °C/min during the measurement. The melting curves were drawn after normalizing the absorbance value at 260 nm (A_{260}) as follows:

$$\text{Normalized } A_{260} = \{A_{260}(t\,^\circ\text{C}) - A_{260}(4\,^\circ\text{C})\} \\ /\{A_{260}(60\,^\circ\text{C}) - A_{260}(4\,^\circ\text{C})\}.$$

The melting temperature (T_m) was calculated from an inflection point of a melting curve using a T_m analysis software LabSolutions (Shimadzu) with a 17-point adaptive smoothing program. The T_m values were determined from the average of three independent measurements.

CD Spectroscopy

The samples were prepared as described above. CD spectra were measured by a JASCO J-820 spectropolarimeter with 10-time accumulation using a path length of 0.3 cm at 5 °C. The obtained spectra were smoothed using a simple moving average smoothing program.

Metal-Dependent Switching of the Hybridization Partner of dcaU-Containing DNA Strands

For native polyacrylamide gel electrophoresis (PAGE) analysis, FAM-labeled strand **8A′** was used. The samples were prepared by combining strands **7**, **8**, and **8A′** (0.5 μM each) in 10 mM HEPES–NaOH buffer (pH 7.0) containing 100 mM NaCl. The solutions were annealed (from 85 °C to 4 °C, – 1.0 °C/min) in the absence and presence of varying amounts of Gd^{III} ions and EDTA. Native PAGE analysis was conducted with 20% polyacrylamide gel in a cool incubator (4 °C). The bands were detected by FAM fluorescence and the yield of each product was determined by comparing the intensities of the two bands observed in the same lane.

Gd^{III}-Responsive Exchange of the Hybridization Partner of dcaU-Containing DNA Strands under Isothermal Conditions

For time-course fluorescence analysis, strand **8A′** containing a fluorescent 2-aminopurine (AP) instead of a natural adenine was used. The samples were prepared by combining strands **7**, **8**, and **8A′** (0.5 μM each) in 10 mM HEPES–NaOH buffer (pH 7.0) containing 100 mM NaCl. After the samples were annealed (from 85 °C to 20 °C, – 1.0 °C/min), the fluorescence intensity of AP ($\lambda_{ex} = 303$ nm, $\lambda_{em} = 371$ nm) was measured in 30-s intervals at a constant temperature of 20 °C. After 5 min, Gd^{III} (1 equiv.) was added to induce the strand exchange. For the standard toehold-mediated strand exchange, strand **8A** was added to the preannealed mixture of **7T** and **8A′**. For a control experiment using natural T bases instead of **dcaU**, Gd^{III} was added to the mixture of **7T**, **8A′**, and **8T**. After the solution was further incubated for 15 min, EDTA (5 equiv.) was added to remove Gd^{III} ions. The fluorescence intensity (I_{obs}) was normalized based on the emission intensity values of duplex **7**·**8A′** or **7T**·**8A′** (I_{dsDNA}) and single strand **8A′** (I_{ssDNA}) as follows:

$$\text{Normalized intensity} = (I_{obs} - I_{dsDNA})/(I_{ssDNA} - I_{dsDNA}).$$

Reversible Switching of the Hybridization Partner of dcaU-Containing DNA Strands under Isothermal Conditions

The sample was prepared as described above. After annealed (from 85 °C to 20 °C, – 1.0 °C/min), the sample was subjected to time-course fluorescence analysis ($\lambda_{ex} = 303$ nm, $\lambda_{em} = 371$ nm) with 30-s intervals at a constant temperature of 20 °C. After 5 min, Gd^{III} (1 equiv.), EDTA (5 equiv.), Gd^{III} (5 equiv.), EDTA (25 equiv.), Gd^{III} (25 equiv.), and EDTA (125 equiv.) were sequentially added at the defined time points. The fluorescence intensity was normalized as described above.

References

1. Takezawa Y, Nishiyama K, Mashima T, Katahira M, Shionoya M (2015) Chem Eur J 21:14713–14716
2. Nishiyama K, Takezawa Y, Shionoya M (2016) Inorg Chim Acta 452:176–180
3. Nishiyama K, Mori K, Takezawa Y, Shionoya M (2021) Chem Commun 57:2487–2490
4. Takezawa Y, Suzuki A, Nakaya M, Nishiyama K, Shionoya M (2020) J Am Chem Soc 142:21640–21644
5. Corradi AB, Palmieri CG, Nardelli M, Pellinghelli MA, Tani MEV (1973) J Chem Soc Dalton Trans 655–658
6. Schmitt W, Jordan PA, Henderson RK, Moore GR, Anson CE, Powell AK (2002) Coord Chem Rev 228:115–126
7. Walters MA, Vapnyar V, Bolour A, Incarvito C, Rheingold AL (2003) Polyhedron 22:941–946
8. Tribet M, Covelo B, Choquesillo-Lazarte D, González-Pérez JM, Castiñeiras A, Niclos-Gutiérrez J (2003) Inorg Chem Commun 6:343–345
9. Ren Y-P, Long L-S, Mao B-W, Yuan Y-Z, Huang R-B, Zheng L-S (2003) Angew Chem Int Ed 42:532–535
10. Zhang Q-Z, He X, Yu Y-Q, Chen S-M, Lu C-Z, Anorg Z (2005) Allg Chem 631:798–802
11. Thompson LC (1962) Inorg Chem 1:490–493
12. Tsien RY (1980) Biochemistry 19:2396–2404
13. Bünzli J-CG (2014) J Coord Chem 67:3706–3733
14. Martinez-Gomez NC, Vu HN, Skovran E (2016) Inorg Chem 55:10083–10089
15. Gourdain S, Petermann C, Harakat D, Clivio P (2010) Nucleosides Nucleotides Nucleic Acids 29:542–546
16. Sandmann N, Defayay D, Hepp A, Müller J (2019) J Inorg Biochem 191:85–93
17. Hu L, Takezawa Y, Shionoya M (2022) Chem Sci 13:3977–3983
18. Jean JM, Hall KB (2001) Proc Natl Acad Sci USA 98:37–41
19. Takezawa Y, Shionoya M (2012) Acc Chem Res 45:2066–2076
20. Takezawa Y, Müller J, Shionoya M (2017) Chem Lett 46:622–633
21. Naskar S, Guha R, Müller J (2020) Angew Chem Int Ed 59:1397–1406
22. Duffy K, Arangundy-Franklin S, Holliger P (2020) BMC Biol 18:112
23. Tanaka K, Tengeiji A, Kato T, Toyama N, Shiro M, Shionoya M (2002) J Am Chem Soc 124:12494–12498
24. Tanaka K, Tengeiji A, Kato T, Toyama N, Shionoya M (2003) Science 299:1212–1213
25. Takezawa Y, Nakama T, Shionoya M (2019) J Am Chem Soc 141:19342–19350
26. Nakama T, Takezawa Y, Sasaki D, Shionoya M (2020) J Am Chem Soc 142:10153–10162

Chapter 4
Metal-Responsive DNA Tweezers Driven by Base Pair Switching of 5-Hydroxyuracil Nucleobases

Abstract Controlling DNA nanomachines in response to metal ions will broadly extend their functions and rate of operation. In this chapter, I operated DNA tweezers by metal-responsive base pair switching of $\mathbf{U^{OH}}$ bases. A pair of DNA tweezers was designed to be closed by hybridization of a $\mathbf{U^{OH}}$-containing closing strand through $\mathbf{U^{OH}}$–A pairing. The addition of Gd^{III} ions was expected to destabilize the $\mathbf{U^{OH}}$–A pairs and release the closing strand with the $\mathbf{U^{OH}}$–Gd^{III}–$\mathbf{U^{OH}}$ complexes, opening the tweezers. Native polyacrylamide gel electrophoresis revealed that the $\mathbf{U^{OH}}$-modified tweezers were predominantly closed (66%) in the absence of Gd^{III} ions. On the other hand, the addition of Gd^{III} released the $\mathbf{U^{OH}}$-containing closing strand along with $\mathbf{U^{OH}}$–Gd^{III}–$\mathbf{U^{OH}}$ complexes, opening 83% of the tweezers. The Gd^{III}-dependent structural conversion was also proven by fluorescence analysis using tweezers with a pair of fluorophore and quencher. Furthermore, the $\mathbf{U^{OH}}$-modified tweezers were reversibly operated by the addition and removal of Gd^{III} ions under isothermal conditions. These results demonstrated that the base pair switching of $\mathbf{U^{OH}}$ would be widely applicable to metal-responsive operation of DNA molecular machines.

Keywords Molecular machine · DNA tweezers · Base pair switching · 5-hydroxyuracil · Reversible switching

4.1 Introduction

A variety of nanostructures with a well-defined shape and size have been constructed based on the sequence-dependent self-assembly of DNA molecules [1–5]. Furthermore, dynamic control of the DNA assemblies has been achieved based on strand displacement reactions (SDRs) [6, 7], enabling the development of DNA-based molecular machines and nanodevices [8–10]. Due to its highly predictable secondary structures and programmable hybridization kinetics, DNA has been recognized as a highly suitable material for constructing artificial nanomachines.

K. Mori, *Metal-Responsive Base Pair Switching of Ligand-type Uracil Nucleobases*, Springer Theses, https://doi.org/10.1007/978-981-99-9400-7_4

DNA tweezers have been one of the most typical DNA-based nanodevices, showing sharp switching of their structures between closed and open forms. The closed state of the tweezers is shown in Fig. 4.1a. Four DNA strands are assembled to form a tweezer-like structure with two duplex arms and a hinge. The tweezers are closed in the presence of a strand complementary to the two arms, called "closing strand". Without the closing strand, the two arms are no longer fixed and the open tweezers are formed. In the pioneering work by Yurke et al., they operated the tweezers based on toehold-mediated SDRs (Fig. 4.1b) [11]. A toehold region (8 nt) was introduced into the 3′ terminus of the closing strand. By the addition of an input strand (input A), the closing strand is dissociated from the tweezers via toehold-mediated SDRs, resulting in the structural conversion to the open form. The operation of the tweezers was reversible by adding the closing strand (input B) again, and the repeated operation was also demonstrated for several cycles. A robust type of the DNA tweezers was also developed using double-crossover (DX) tile motifs [12]. The structural change of the DX tile-based DNA tweezers has been applied for controlling dynamic functions. For example, it was demonstrated that the target proteins can be captured by the tweezers whose arms are modified with aptamer sequences (Fig. 4.1c) [13]. Furthermore, regulation of enzyme functions was also achieved by introducing a pair of an enzyme and its cofactor at the termini of the tweezers (Fig. 4.1d) [14]. Hence, the DNA tweezers have been an important DNA-based structural motif that exhibits efficiently switchable configurations and broadly applicable functions.

As represented by the DNA tweezers, it has been demonstrated that DNA-based molecular machines can be operated through sequence-dependent hybridization of oligonucleotides. With a view to expanding the applicability of DNA nanomachines, researchers have been working on controlling them in response to external stimuli instead of the oligonucleotide binding [10, 15]. In particular, DNA tweezers which respond to external stimuli have been actively studied. For instance, the tweezers containing azobenzene-modified closing strands can be operated by light irradiation (Fig. 4.2a) [16, 17]. Because the azobenzene-modified strand can form stable DNA duplexes only when the azobenzene takes a *trans* form, the binding of the closing strand can be controlled based on the photoisomerization. As a result, UV light irradiation triggered the opening operation driven by the isomerization of the azobenzenes to the *cis* form. In contrast, the closed state was restored under visible light due to the regeneration of the *trans* form. pH-responsive control of the tweezers has also been achieved by incorporating i-motif structures into the arms (Fig. 4.2b) [18]. The i-motif is a quadruplex formed by stacked pairs of a protonated cytosine (C^+) and a neutral cytosine (C) bases [19]. Accordingly, the i-motif can be stably formed only under acidic conditions. While the closed tweezers are predominantly formed at neutral pH, the i-motif structures were constructed in the tweezer arms at pH 5.2. As a result, the closing strand was dissociated from the tweezers, yielding the open form.

Metal ions show unique properties such as reversible complexation and ligand-selective binding. Therefore, metal-dependent control of DNA molecular machines would allow for reversible and programmable operation, greatly expanding the scope

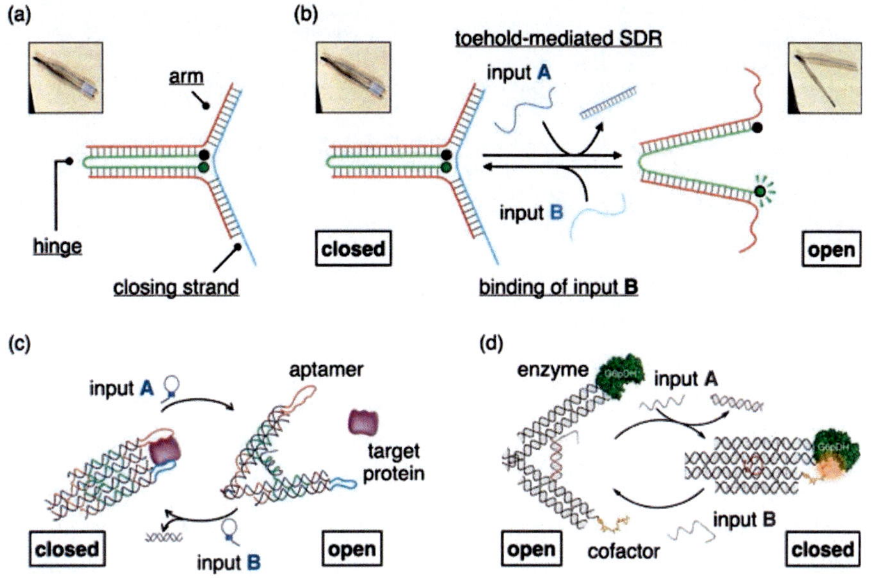

Fig. 4.1 Previous studies on DNA tweezers. **a** A closed form of the DNA tweezers. Adapted with permission from Ref. [17]. Copyright 2014 American Chemical Society. **b** Reversible operation of the DNA tweezers by toehold-mediated strand displacement reactions (SDRs). Adapted with permission from Ref. [17]. Copyright 2014 American Chemical Society. **c** Capture of target proteins by DNA tile-based tweezers modified with aptamer sequences. Adapted with permission from Ref. [13]. Copyright 2012 American Chemical Society. **d** Enzyme nanoreactor regulated by DNA tweezers. Adapted with permission from Ref. [5]. Copyright 2019 American Chemical Society

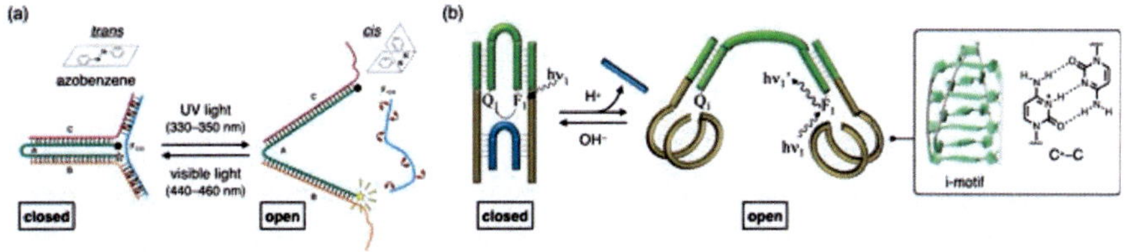

Fig. 4.2 Previous studies on external stimuli-responsive DNA tweezers. **a** Light-responsive DNA tweezers driven by the photoisomerization of azobenzenes. Adapted with permission from Ref. [17]. Copyright 2014 American Chemical Society. **b** pH-responsive DNA tweezers based on the formation of i-motif structures in the tweezer arms. Adapted with permission from Refs. [10] and [18]. Copyright 2021 Shiliang He et al. and 2009 American Chemical Society

of dynamic DNA nanotechnology. For this purpose, the base pair switching of the 5-modified uracil bases is useful. As demonstrated in Chap. 2, the 5-hydroxyuracil ($\mathbf{U^{OH}}$) base can form not only hydrogen-bonded $\mathbf{U^{OH}}$–A base pairs, but also metal-mediated $\mathbf{U^{OH}}$–Gd^{III}–$\mathbf{U^{OH}}$ pairs. It was expected that the base pair switching of $\mathbf{U^{OH}}$ can be applied to metal-dependent conversion of DNA structures between two distinct assembly states. In this study, I aimed to prove the general applicability

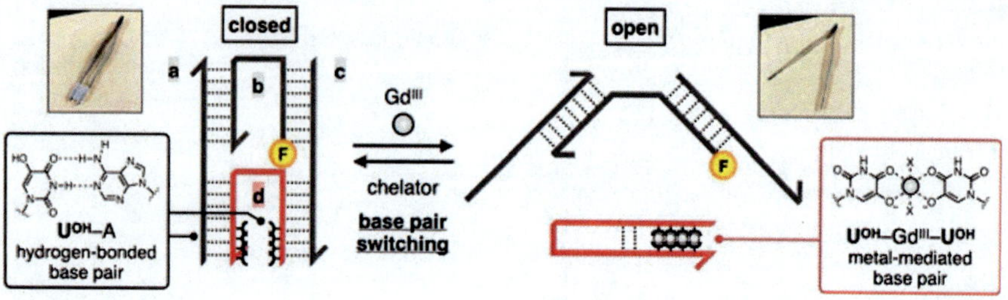

Fig. 4.3 Design of Gd^{III}-responsive DNA tweezers containing **U^{OH}** bases. Strand **d** contains multiple **U^{OH}** bases at both termini

of the base pair switching and developed metal-responsive DNA tweezers with the **U^{OH}** bases. The metal-responsive DNA tweezers were designed by introducing **U^{OH}** bases into the two arms (Fig. 4.3). The closed form of the tweezers can be stably constructed through the **U^{OH}**–A pairing. In contrast, it was expected that the addition of Gd^{III} ions would destabilize the **U^{OH}**–A pairs and the intrastrand **U^{OH}**–Gd^{III}–**U^{OH}** complexation would lead to the release of the **U^{OH}**-containing closing strand, yielding the open state.

In this chapter, first I explain the design strategy of the **U^{OH}**-containing DNA tweezers (Sect. 4.2). Then I confirmed the formation the tweezer structures and optimized the sequences and solution conditions for the efficient structural conversion in response to Gd^{III} ions (Sect. 4.3). Finally, Gd^{III}-responsive reversible operation of the **U^{OH}**-containing tweezers was demonstrated under isothermal conditions (Sect. 4.4).

4.2 Design of Metal-Responsive DNA Tweezers Containing 5-Hydroxyuracil Bases

Metal-responsive DNA tweezers were developed by modifying the reported tweezer structures with **U^{OH}** bases as Gd^{III}-responsive units. The base sequences of the **U^{OH}**-containing tweezers, named as **U^{OH}**-tweezers, were designed based on the pioneering work by Yurke et al. [11]. They constructed DNA tweezers with four strands **a**, **b**, **c**, and **d**, and reversibly operated them through toehold-mediated strand displacement reactions (SDRs) (Fig. 4.4a). In Chap. 2, I demonstrated that the Gd^{III}-triggered SDRs efficiently proceeded with DNA strands containing four consecutive **U^{OH}** bases at the termini. According to this result, four **U^{OH}** bases were introduced at both termini of closing strand **d** instead of the toehold sequences (Fig. 4.4b). It was expected that **U^{OH}**-tweezers can be closed by the binding of strand **d** through **U^{OH}**–A pairing. In the presence of Gd^{III} ions, the **U^{OH}**–A pairs would be destabilized and intrastrand **U^{OH}**–Gd^{III}–**U^{OH}** pairs can be formed in strand **d**. As a result, strand **d** would be dissociated from the tweezers, yielding the open structure. Furthermore, the Gd^{III}-responsive base pair switching of **U^{OH}** is reversible by removing Gd^{III} ions with chelating agents, as described in Chap. 2. Therefore, it was also expected that the

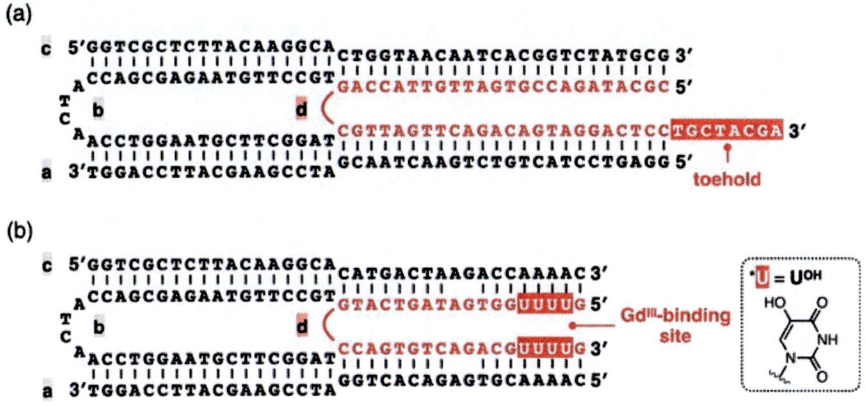

Fig. 4.4 Sequence design of Gd^{III}-responsive **U^OH**-tweezers. **a** Base sequences of the original DNA tweezers reported by Yurke et al. A toehold sequence is highlighted with a red box. **b** Base sequences of **U^OH**-tweezers containing four **U^OH** bases at both termini of the arms. U stands for the **U^OH** bases as Gd^{III}-binding sites, which are also highlighted with red boxes

closed tweezers can be regenerated through the removal of Gd^{III} ions by chelator molecules.

For the formation of the closed tweezers, stable binding of closing strand **d** to the tweezer structures is required. On the other hand, strand **d** should be dissociated from the tweezers in the presence of Gd^{III} ions to give the open form. In order to realize the Gd^{III}-dependent binding of strand **d**, the hybridization strength of strand **d** should be carefully designed by arranging the base sequences. Accordingly, three base sequences were designed (Design 1–3) (Fig. 4.5). Considering that the Gd^{III}-responsive strand exchange efficiently proceeded with 17-mer DNA strands as shown in Chap. 2, the length of the tweezer arms was fixed to 17 base pairs (bp) throughout the three designs. The two bases in the middle of strand **d** (G and C, highlighted in yellow in Fig. 4.5a) were not expected to form stable base pairs with the tweezer arms because the kinking of the strand causes severe structural distortion. In Design 1, strand **d** binds to the arms of the tweezers by the formation of full-match duplexes **a1·d** and **c1·d** (Fig. 4.5a). The Gd^{III}-dependent structural conversion of the tweezers was modeled by predicting thermal stability of the tweezer structures using NUPACK software [20, 21]. For the calculation, only the arm region was modeled and the **U^OH**–A pairs were replaced by natural T–A pairs. To model the Gd^{III}-containing state, the destabilization of the **U^OH**–A pairs should be taken into consideration. As shown in Chap. 2, the melting temperature (T_m) of the duplex with four **U^OH**–A pairs in the presence of Gd^{III} ions (T_m = 38.9 °C) was similar to that of the duplex with T–T mismatches (37.7 °C). Therefore, the Gd^{III}-containing state was modeled by introducing the T–T mismatches instead of the **U^OH**–A pairs. Figure 4.5b shows the predicted melting profile of the tweezer arms. As a result, it was suggested that the addition of Gd^{III} ions lowers the stability of the arm duplexes. However, the T_m value of the arms was approximately 45 °C even under the Gd^{III}-containing condition. This result suggests that the tweezers cannot be opened at 25 °C, which is a standard condition for the tweezers operation. According to this estimation, I have designed two more sequences (Design 2 and 3), in which one and two mismatches were introduced into the tweezer arms, respectively (Fig. 4.5a). It was expected that the duplex stability is lowered by the incorporation of mismatches and the dissociation of closing strand **d** is promoted in the presence of Gd^{III} ions. The modeling results revealed that the thermal stability of the arm duplexes was decreased by introducing the mismatches (Fig. 4.5, mismatches are highlighted in blue in Fig. 4.5a). In particular, Design 3 gave the T_m value of approximately 30 °C under the Gd^{III}-containing condition, indicating that the opening of the tweezers became more probable. It should be noted that this calculation is only an approximation and the actual stability of the tweezers would be affected by various factors such as salt concentrations and amounts of Gd^{III} ions. Therefore, I decided to experimentally optimize the sequences and conditions for the operation of the DNA tweezers based on these three sequence designs.

The designed $\mathbf{U^{OH}}$-containing strand **d** was synthesized by an automated synthesizer using $\mathbf{U^{OH}}$ phosphoramidite in the same manner as described in Chap. 2, Sect. 2.2. After cleavage from the solid support and deprotection, the synthesized

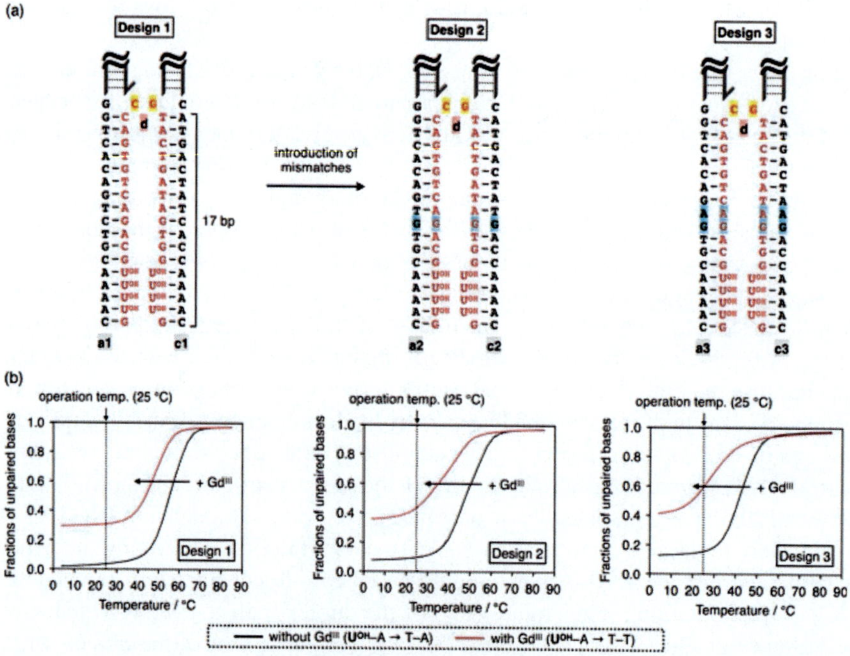

Fig. 4.5 Sequence designs of the arms in $\mathbf{U^{OH}}$-tweezers. **a** Base sequences of the tweezers. While Design 1 contains full-match duplexes, one and two mismatches were introduced in Design 2 and 3, respectively. The mismatch bases are highlighted with blue boxes. Two bases highlighted in yellow are not expected to form base pairs with the arms. **b** Predicted melting profiles of the arm duplexes in each design. The melting curves were calculated by NUPACK software [20, 21]. For the analysis, the $\mathbf{U^{OH}}$–A pairs were replaced by T–A pairs (black lines) or T–T mismatches (red lines)

strand was purified by denaturing PAGE. The obtained sample was further purified by size-exclusion gel chromatography and isopropanol precipitation. Finally, strand **d** was characterized by MALDI-TOF mass spectrometry (see Sect. 4.6).

In summary, $\mathbf{U^{OH}}$-tweezers were designed based on the base sequences reported in the original work of DNA tweezers [11]. Four $\mathbf{U^{OH}}$ bases were introduced at both termini of the arms as Gd^{III}-binding sites to induce efficient release of strand **d** in the presence of Gd^{III} ions. With a view to optimizing the sequences, three sequences with different numbers of mismatches in the tweezer arms were designed. It was expected that comparison of the three designs under variable experimental conditions such as buffer conditions and Gd^{III} concentrations would give the optimized design of $\mathbf{U^{OH}}$-tweezers.

4.3 Formation and Metal-Dependent Structural Conversion of DNA Tweezers Containing 5-Hydroxyuracil Bases

As described in Sect. 4.2, Gd^{III}-responsive DNA tweezers ($\mathbf{U^{OH}}$-tweezers) were designed by introducing four $\mathbf{U^{OH}}$ bases at both termini of closing strand **d**. Herein, $\mathbf{U^{OH}}$-tweezers have been constructed and the Gd^{III}-dependent structural conversion was studied.

DNA strands used in this section is summarized in Table 4.1. The termini of strand **b** were labeled with a FAM and Dabcyl ($\mathbf{b_F}$ and $\mathbf{b_{FD}}$) for the characterization of the DNA tweezers by gel electrophoresis and fluorescence spectroscopy (Fig. 4.6a). Strand $\mathbf{d_T}$, which contains natural T bases instead of $\mathbf{U^{OH}}$, was also prepared to construct DNA tweezers containing T–A pairs instead of $\mathbf{U^{OH}}$–A pairs (T-tweezers) for control experiments (Fig. 4.6b).

Table 4.1 Sequences of DNA strands used in Chap. 4

Name	Sequence (5′ to 3′)
a1	CAC ATT GTC TGA CAC TGG ATC CGA AGC ATT CCA GGT
a2	CAA AAC GTG TGA CAC TGG ATC CGA AGC ATT CCA GGT
a3	CAA AAC GTG AGA CAC TGG ATC CGA AGC ATT CCA GGT
b_F	FAM—TGC CTT GTA AGA GCG ACC ATC AAC CTG GAA TGC TTC GGA T
b_{FD}	FAM—TGC CTT GTA AGA GCG ACC ATC AAC CTG GAA TGC TTC GGA T
c1	GGT CGC TCT TAC AAG GCA CAT GAC TAT CAC CAA AAC
c2	GGT CGC TCT TAC AAG GCA CAT GAC TAT GAC CAA AAC
c3	GGT CGC TCT TAC AAG GCA CAT GAC TAA GAC CAA AAC
d	G$\mathbf{U^{OH}U^{OH}}$ $\mathbf{U^{OH}U^{OH}}$G GTG TTA GTC ATG CCA GTG TCA CAC G$\mathbf{U^{OH}U^{OH}}$ $\mathbf{U^{OH}U^{OH}}$G
d_T	G**TT TT**G GTG TTA GTC ATG CCA GTG TCA CAC G**TT TT**G

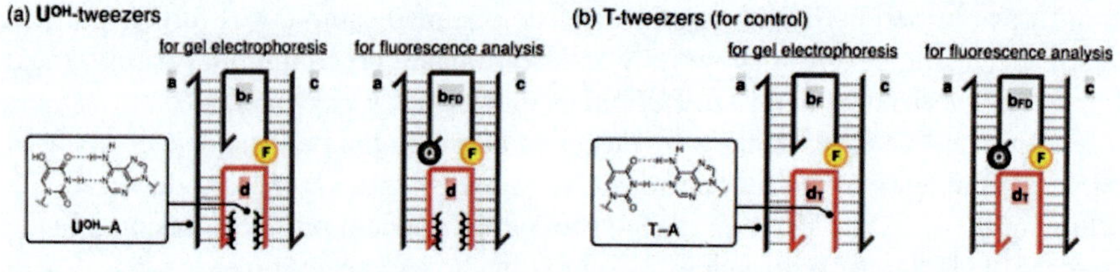

Fig. 4.6 Designs of **a** $\mathbf{U^{OH}}$-tweezers containing four $\mathbf{U^{OH}}$ bases at both termini of the tweezer arms and **b** T-tweezers containing natural T bases instead of $\mathbf{U^{OH}}$. F: FAM, Q: Dabcyl

4.3.1 Optimization of the Base Sequences of the Tweezer Arms

Based on the three designs shown in Fig. 4.5, the base sequences of the duplex arms were optimized. Formation of the DNA tweezers was first confirmed by T-tweezers using strand $\mathbf{d_T}$, which contains natural T bases instead of $\mathbf{U^{OH}}$ (Fig. 4.7a). Three DNA strands **a1**, $\mathbf{b_F}$, and **c1** were mixed in the absence and presence of strand $\mathbf{d_T}$ to form the open and closed structures of T-tweezers, respectively. Figure 4.7b shows the results of native polyacrylamide gel electrophoresis (PAGE) analysis. In the absence of strand $\mathbf{d_T}$, a single band was observed, suggesting the formation of the open state of the tweezers. In contrast, the addition of strand $\mathbf{d_T}$ gave a band with lower mobility, confirming that T-tweezers were closed by the hybridization of strand $\mathbf{d_T}$. Another band observed in the presence of strand $\mathbf{d_T}$ can be a dimer structure in which two DNA tweezers are connected by binding of strand $\mathbf{b_F}$ (Fig. 4.7c). Such a dimer structure was also reported in the previous study [11].

Next, Gd^{III}-dependent structural conversion of $\mathbf{U^{OH}}$-tweezers was studied using Design 1 with the full-match arm duplexes. $\mathbf{U^{OH}}$-containing strand **d** and a set of the other three strands (**a1**, $\mathbf{b_F}$, and **c1**) were mixed and annealed in the absence and presence of Gd^{III} ions (Fig. 4.8). Under Gd^{III}-free conditions, the band of the closed tweezers was clearly observed in native PAGE analysis, indicating quantitative formation of the closed structure. However, the gel pattern was not significantly changed even in the presence of Gd^{III} ions (4 equiv.). These results suggest that

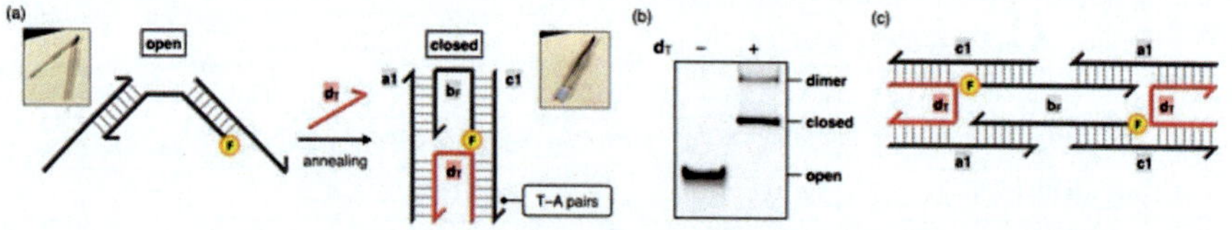

Fig. 4.7 Formation of the open and closed structures of T-tweezers without and with strand $\mathbf{d_T}$. **a** Scheme. F: FAM. **b** Native PAGE analysis of the tweezers in the absence and presence of strand $\mathbf{d_T}$. [each DNA] = 2.0 μM in 10 mM HEPES–NaOH buffer (pH 8.0) containing 100 mM NaCl. 15% gel at 25 °C. **c** A dimeric structure of T-tweezers

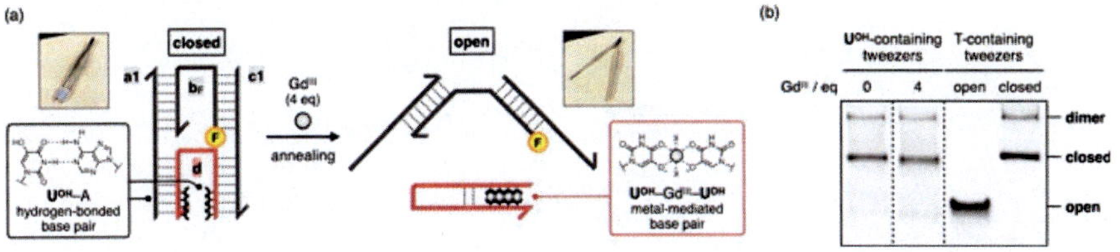

Fig. 4.8 Gd^{III}-dependent structural conversion of $\mathbf{U^{OH}}$-tweezers (Design 1). **a** Scheme. F: FAM. **b** Native PAGE analysis of the tweezers in the absence and presence of Gd^{III} ions (4 equiv.). [each DNA] = 2.0 μM, [Gd^{III}]/[each DNA] = 0 or 4 in 10 mM HEPES–NaOH buffer (pH 8.0) containing 100 mM NaCl. 15% gel at 25 °C

$\mathbf{U^{OH}}$-tweezers of Design 1 cannot be opened by adding Gd^{III} ions. This is possibly because the binding of strand **d** was too strong in Design 1.

Based on the result of Design 1, it was expected that the thermal stability of the arm duplexes should be lowered to induce Gd^{III}-dependent dissociation of strand **d** and to yield the open tweezers. Accordingly, Design 2 and 3 containing one and two mismatches in the arms, respectively, were tested (Fig. 4.9a). In Design 2, the tweezers were closed with or without Gd^{III} ions (Fig. 4.9b). Therefore, it was indicated that the introduction of one mismatch was not enough to induce the Gd^{III}-dependent dissociation of the arm duplexes. In contrast, native PAGE analysis of Design 3 gave a smeared band in the presence of Gd^{III} ions (Fig. 4.9c). This result suggests that the binding of strand **d** to the tweezers was weakened due to the $\mathbf{U^{OH}}$–Gd^{III}–$\mathbf{U^{OH}}$ complexation, resulting in gradual dissociation of strand **d** during the electrophoresis. The results show that introducing two consecutive mismatches in each arm of the tweezers can promote Gd^{III}-dependent structural conversion to the open form.

In summary, the Gd^{III}-dependent structural conversion of $\mathbf{U^{OH}}$-tweezers was observed only in Design 3. Accordingly, Design 3 was adopted as the optimized sequence and used for further experiments.

4.3.2 *Evaluation of the Structural Conversion by Fluorescence Measurement*

The "closed" and "open" configurations of the tweezers were further confirmed by fluorescence spectroscopy using DNA strands with the optimized sequences (Design 3). For this purpose, strand $\mathbf{b_{FD}}$ containing a FAM fluorophore and a Dabcyl quencher at each terminus was used. In the closed form of the tweezers, the FAM and Dabcyl are located in close proximity, and thus the emission from the FAM would be quenched. On the other hand, the open form of the tweezers makes the two moieties distant from each other, resulting in the recovered fluorescence. T-tweezers were first subjected to the fluorescence analysis to confirm that the structural conversion of the tweezers can be evaluated by the fluorescence changes. As shown in Fig. 4.10, the fluorescence

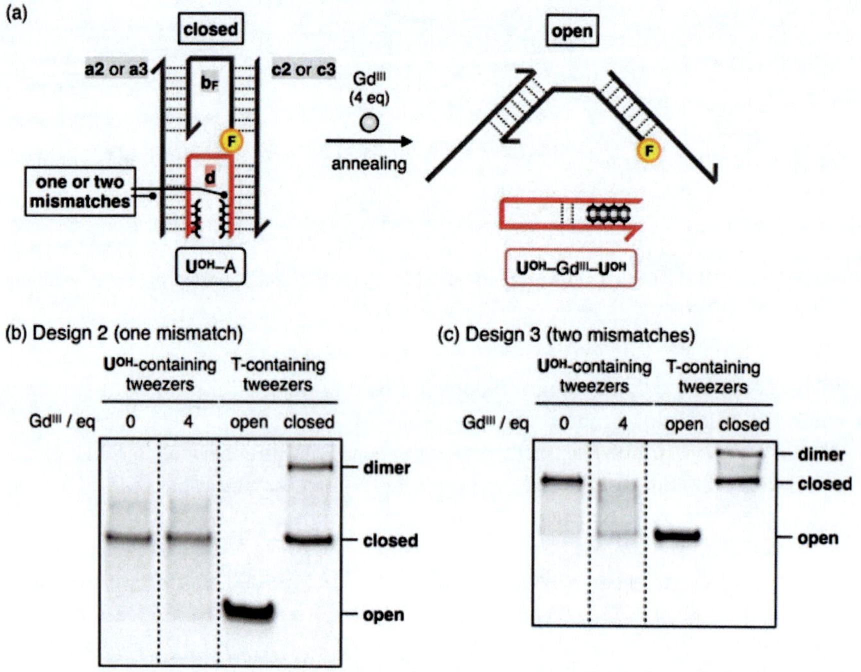

Fig. 4.9 Gd^{III}-dependent structural conversion of $\mathbf{U^{OH}}$-tweezers (Design 2 and 3). **a** Scheme. F: FAM. **b**, **c** Native PAGE analysis of the tweezers in the absence and presence of Gd^{III} ions (4 equiv.). **b** Design 2, **c** Design 3. [each DNA] = 2.0 μM, $[Gd^{III}]$/[each DNA] = 0 or 4 in 10 mM HEPES–NaOH buffer (pH 8.0) containing 100 mM NaCl. 15% gel at 25 °C

intensity was significantly decreased in the presence of strand $\mathbf{d_T}$. This result suggests the formation of the closed tweezers through T–A pairing in the tweezer arms.

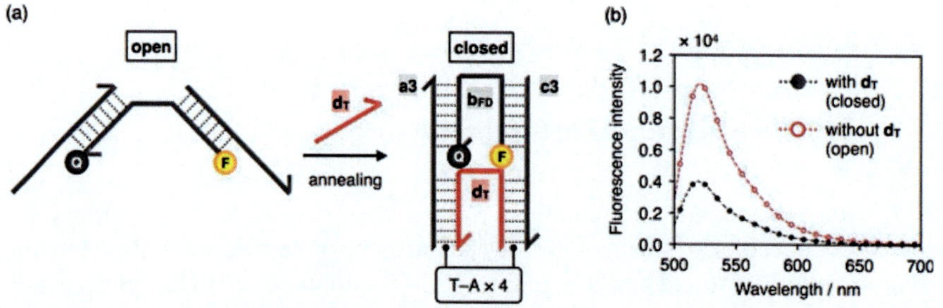

Fig. 4.10 Fluorescence analysis of T-tweezers containing natural T bases instead of $\mathbf{U^{OH}}$. **a** Scheme. F: FAM, Q: Dabcyl. **b** Fluorescence spectra of the tweezers in the absence and presence of strand $\mathbf{d_T}$. [each DNA] = 2.0 μM, in 10 mM HEPES–NaOH buffer (pH 8.0) containing 100 mM NaCl. λ_{ex} = 495 nm, 25 °C

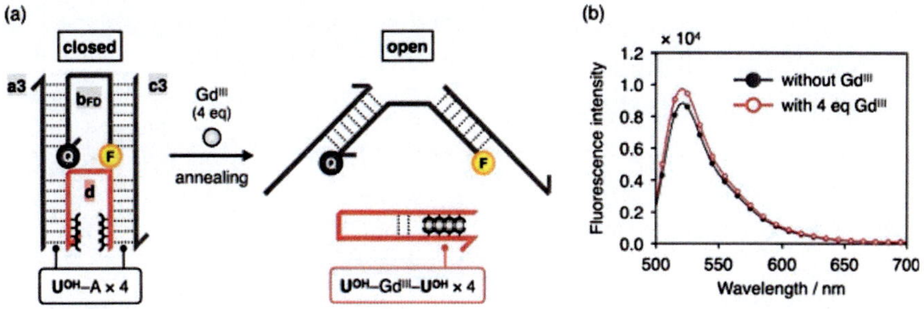

Fig. 4.11 Fluorescence-based analysis of **U^{OH}**-tweezers (Design 3) in the absence and presence of Gd^{III} ions. **a** Scheme. F: FAM, Q: Dabcyl. **b** Fluorescence spectra of the tweezers in the absence and presence of Gd^{III} ions. [each DNA] = 2.0 μM, [Gd^{III}]/[each DNA] = 0 or 4 in 10 mM HEPES–NaOH buffer (pH 8.0) containing 100 mM NaCl. λ_{ex} = 495 nm, 25 °C

Subsequently, the Gd^{III}-dependent structural conversion of **U^{OH}**-tweezers (Design 3) was evaluated by fluorescence measurement (Fig. 4.11a). In the presence of 4 equiv. of Gd^{III} ions, the emission intensity of the FAM was almost comparable with that of the tweezers without strand **d** (Fig. 4.11b, red line). Therefore, it was suggested that the tweezers took the open structures due to the dissociation of strand **d** from the tweezers. However, the fluorescence from the FAM was only slightly quenched even under Gd^{III}-free conditions (Fig. 4.11b, black line). This result indicates that strand **d** cannot stably bind to the tweezers to close their structures under this condition.

The fluorescence spectra of **U^{OH}**-tweezers shown in Fig. 4.11b were compared with those of T-tweezers to estimate the yields of the open and closed forms (Fig. 4.12). Since T-tweezers showed quantitative formation of the closed and open states depending on the presence of strand $\mathbf{d_T}$ (Fig. 4.9c), the fluorescence intensity of T-tweezers can be used as a reference of each state. Accordingly, the fluorescence intensities of **U^{OH}**-tweezers were normalized by those of T-tweezers to calculate the yield of the open form. The results revealed that 72% of the tweezers open without Gd^{III} ions. This result was not consistent with native PAGE analysis of **U^{OH}**-tweezers shown in Fig. 4.9, which suggested the quantitative formation of the closed tweezers in the absence of Gd^{III} ions. This discrepancy may be due to the presence of Mg^{II} ions in TAMg buffer used for native PAGE analysis, which could stabilize the closed state of the DNA tweezers.

In summary, the structural conversion of T-tweezers was confirmed by fluorescence measurement. However, the fluorescence changes of **U^{OH}**-tweezers were much smaller than expected. These results suggest that the Mg^{II} concentration in the tweezers samples should be optimized.

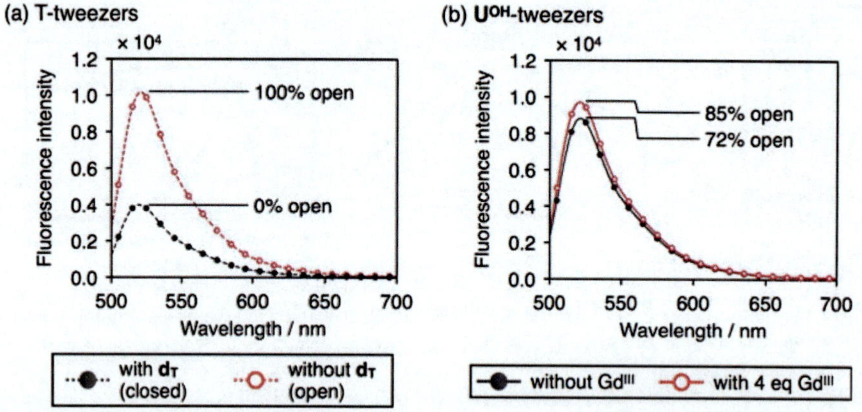

Fig. 4.12 Comparison of the fluorescence spectra of **a** T-tweezers and **b** $\mathbf{U^{OH}}$-tweezers (Design 3). The fluorescence intensity of $\mathbf{U^{OH}}$-tweezers was normalized by that of T-tweezers to calculate the yield of the open form. [each DNA] = 2.0 μM, [Gd^{III}]/[each DNA] = 0 or 4 in 10 mM HEPES–NaOH buffer (pH 8.0) containing 100 mM NaCl. $\lambda_{ex} = 495$ nm, 25 °C

4.3.3 Optimization of Mg^{II} Concentration

For self-assembly of DNA strands, metal cations are necessary to neutralize the negative charges on the phosphate backbones. In particular, divalent Mg^{II} ions remarkably stabilize DNA assemblies. It is expected that the closed form of the DNA tweezers suffers from more electrostatic repulsion than the open form because the two arm duplexes are in close proximity. Therefore, it is worth optimizing the Mg^{II} concentration to stabilize the closed tweezers and to achieve efficient operation of the DNA tweezers. The effects of Mg^{II} ions on the structural conversion of the tweezers were investigated by fluorescence measurement. Figure 4.13 represents the fluorescence spectra of T-tweezers with varying concentrations of Mg^{II} ions. As shown in Fig. 4.13a, the closed state of T-tweezers showed gradual decrease in the fluorescence intensity as the Mg^{II} concentration was increased. This change occurred possibly because (i) the structure of the closed tweezers became more rigid to facilitate the energy transfer between the FAM and Dabcyl or (ii) the formation of the closed state was enhanced. In contrast, only slight intensity change was observed for the open tweezers (Fig. 4.13b). Then the effects of Mg^{II} ions on $\mathbf{U^{OH}}$-tweezers were also evaluated (Fig. 4.14). It was revealed that the fluorescence intensity of $\mathbf{U^{OH}}$-tweezers was gradually diminished by increasing the Mg^{II} concentration both in the absence and presence of Gd^{III} ions. This result suggested that Mg^{II} ions stabilized the closed form of $\mathbf{U^{OH}}$-tweezers by promoting the binding of strand **d** to the tweezers.

The yields of the open tweezers were calculated at various Mg^{II} concentrations by normalizing the fluorescence intensity of $\mathbf{U^{OH}}$-tweezers by that of T-tweezers, in the same manner as Fig. 4.12. Figure 4.15a shows the fluorescence spectra of $\mathbf{U^{OH}}$-tweezers and T-tweezers in the presence of 2 mM Mg^{II} ions as an example. By using the open and closed states of T-tweezers as the references, the yields of the

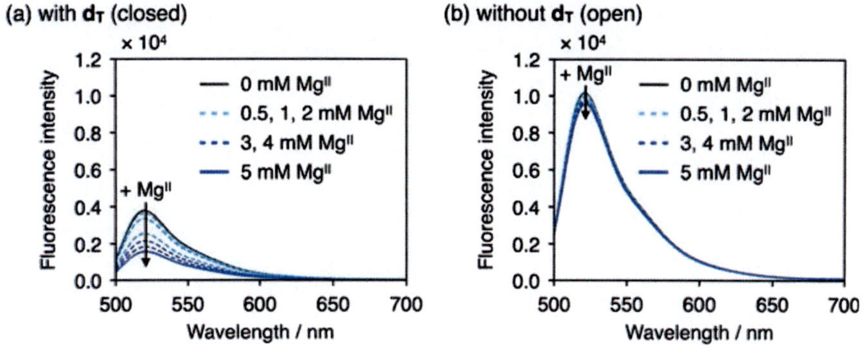

Fig. 4.13 Fluorescence spectra of T-tweezers in the presence of varying concentrations of Mg^{II} ions. **a** With strand $\mathbf{d_T}$, **b** without strand $\mathbf{d_T}$. [$MgCl_2$] = 0 (black solid line), 0.5, 1, 2 (light blue dotted lines), 3, 4 (blue dotted lines), and 5 mM (blue solid line). [each DNA] = 2.0 μM in 10 mM HEPES–NaOH buffer (pH 8.0) containing 100 mM NaCl. λ_{ex} = 495 nm, 25 °C

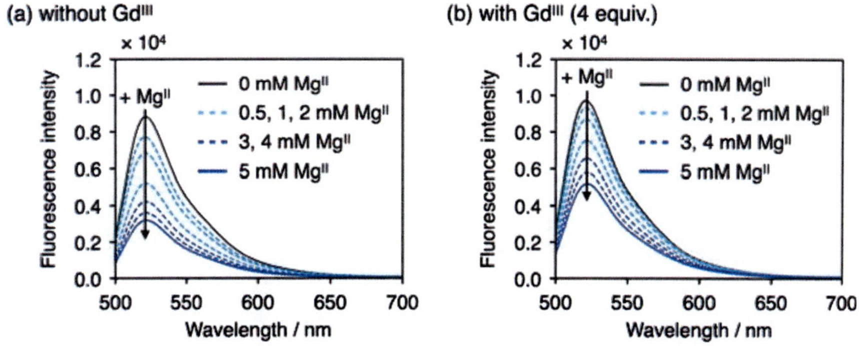

Fig. 4.14 Fluorescence spectra of $\mathbf{U^{OH}}$-tweezers in the presence of varying concentrations of Mg^{II} ions. **a** Without Gd^{III}, **b** with Gd^{III} (4 equiv.). [$MgCl_2$] = 0 (black solid line), 0.5, 1, 2 (light blue dotted lines), 3, 4 (blue dotted lines), and 5 mM (blue solid line). [each DNA] = 2.0 μM, [Gd^{III}]/[each DNA] = 0 or 4 in 10 mM HEPES–NaOH buffer (pH 8.0) containing 100 mM NaCl. λ_{ex} = 495 nm, 25 °C

open tweezers were calculated as 36% and 67% in the absence and presence of Gd^{III} ions, respectively. The same analysis was conducted with varying concentrations of Mg^{II} ions to calculate the yields of the open tweezers (Fig. 4.15b). Furthermore, the efficiency of the Gd^{III}-dependent structural conversion from the closed to open forms of $\mathbf{U^{OH}}$-tweezers, named as "conversion efficiency", was estimated (Fig. 4.15c). The conversion efficiency was defined as the differences in the yields of the open tweezers in the absence and presence of Gd^{III} ions. As a result, the value was maximized at 2 mM and 3 mM Mg^{II} ion concentrations (31% and 30%, respectively).

In conclusion, the fluorescence measurement suggests that the closed state of the tweezers can be stabilized by increasing the concentration of Mg^{II} ions. By comparing the yields of the open tweezers in the absence and presence of Gd^{III} ions,

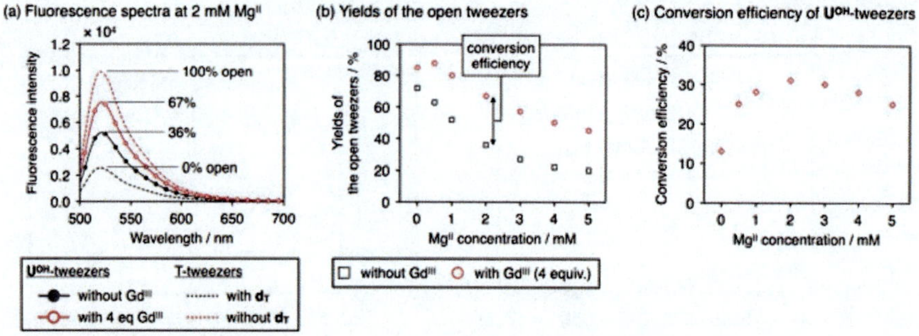

Fig. 4.15 Effects of the Mg^{II} concentration on the yields of the open tweezers and the efficiency of the Gd^{III}-dependent structural conversion. **a** Fluorescence spectra of $\mathbf{U^{OH}}$-tweezers and T-tweezers in the presence of 2 mM Mg^{II} ions. **b** Yields of the open form of $\mathbf{U^{OH}}$-tweezers. The yields were determined by normalizing the fluorescence intensity of $\mathbf{U^{OH}}$-tweezers by that of T-tweezers. **c** Efficiency of the Gd^{III}-dependent structural conversion of $\mathbf{U^{OH}}$-tweezers. [each DNA] = 2.0 μM, $[Gd^{III}]$/[each DNA] = 0 or 4 in 10 mM HEPES–NaOH buffer (pH 8.0) containing 100 mM NaCl. $[MgCl_2]$ = 0, 0.5, 1, 2, 3, 4, or 5 mM. λ_{ex} = 495 nm, 25 °C

the efficiency of the Gd^{III}-dependent structural conversion of $\mathbf{U^{OH}}$-tweezers was evaluated. Based on these analyses, the optimal Mg^{II} concentration was determined as 2 mM and 3 mM Mg^{II} ions.

4.3.4 Optimization of Gd^{III} Concentration

Even under the optimized conditions with 2 mM and 3 mM Mg^{II} ions, the yields of the open tweezers in the presence of Gd^{III} ions (4 equiv.) were only 67 and 57%, respectively. To further enhance the formation of the open structures, the addition of excess Gd^{III} ions was tested (Fig. 4.16). The fluorescence spectra of $\mathbf{U^{OH}}$-tweezers were recorded in the presence of varying equivalents of Gd^{III} ions. Under 2 mM Mg^{II} concentration, 36% of the tweezers was opened in the absence of Gd^{III} ions (Fig. 4.16b). The addition of Gd^{III} ions increased the yield of the open state, resulting in the 83% yield with 12 equiv. of Gd^{III} ions. As a result, the efficiency of the Gd^{III}-dependent structural conversion reached 47%. In the presence of 3 mM Mg^{II} ions, the yield of the open form was changed from 27 to 79% by the addition of Gd^{III} ions (12 equiv.) (Fig. 4.16c). Since the Gd^{III}-dependent conversion with 3 mM Mg^{II} ions (52%) was higher than that with 2 mM Mg^{II} ions (47%), the former condition was selected as an optimized Mg^{II} concentration.

Preferential formation of the open tweezers in the presence of excess Gd^{III} ions was also confirmed by native PAGE (Fig. 4.17). To make the experimental conditions consistent with the fluorescence measurement, TAMg buffer with 3 mM Mg^{II} ions, which was the optimized Mg^{II} concentration for $\mathbf{U^{OH}}$-tweezers, was used for native PAGE. In the absence of Gd^{III} ions, only 34% of the tweezers took the open form. Increasing the amount of Gd^{III} ions gradually improved the yield of open tweezers,

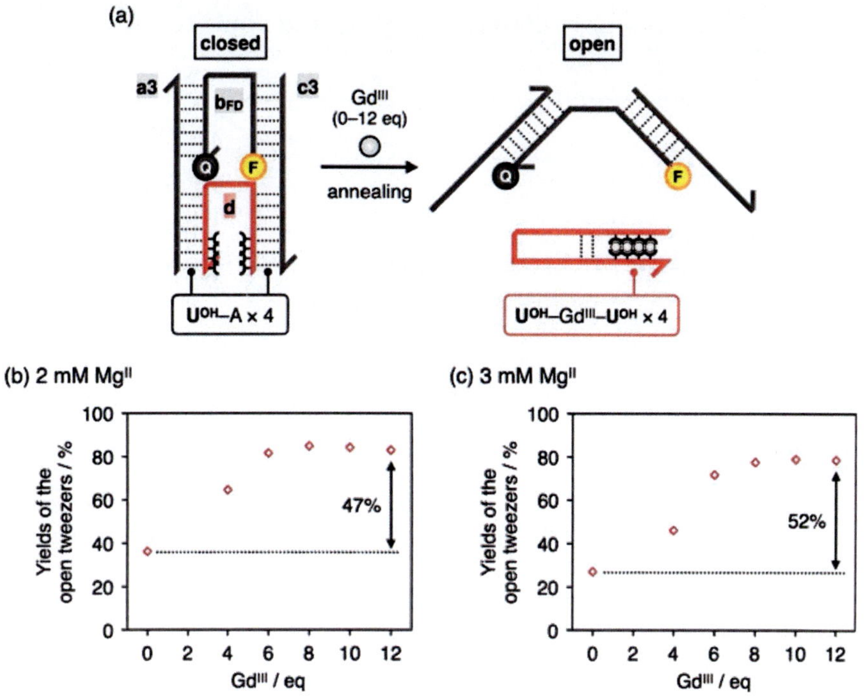

Fig. 4.16 Effects of the Gd^{III} equivalents on the structural conversion of $\mathbf{U^{OH}}$-tweezers. **a** Scheme. F: FAM, Q: Dabcyl. **b**, **c** Plot of the yields of the open tweezers with varying equivalents of Gd^{III} ions. $[MgCl_2]$ = **b** 2 and **c** 3 mM. [each DNA] = 2.0 μM, $[Gd^{III}]$/[each DNA] = 0, 4, 6, 8, 10, and 12 in 10 mM HEPES–NaOH buffer (pH 8.0) containing 100 mM NaCl. λ_{ex} = 495 nm, 25 °C. Yields of the open tweezers under each condition were determined by normalizing the fluorescence intensities with those of T-tweezers

reaching 83% with 12 equiv. of Gd^{III} ions. These yield values were almost consistent with the results of fluorescence analysis shown in Fig. 4.16. Thus, the structural conversion of $\mathbf{U^{OH}}$-tweezers was dramatically enhanced by the addition of excess amount of Gd^{III} ions. This can be because the structural conversion of the tweezers was driven not only by the $\mathbf{U^{OH}}$–$\mathbf{Gd^{III}}$–$\mathbf{U^{OH}}$ base-pair formation, but also by the destabilization of $\mathbf{U^{OH}}$–A pairs. As mentioned in Sect. 2.3 in Chap. 2, stoichiometric or slight excess amount of Gd^{III} ions significantly destabilized duplexes containing $\mathbf{U^{OH}}$–A pairs. Since $\mathbf{U^{OH}}$-tweezers contain eight $\mathbf{U^{OH}}$–A pairs in total, 4 equiv. of Gd^{III} ions were not sufficient and excess Gd^{III} ions were needed to sufficiently destabilize and convert tweezers' structure.

Hence, it was confirmed that the structure of $\mathbf{U^{OH}}$-tweezers can be converted from the closed to open form by the addition of Gd^{III} ions. The base sequences of $\mathbf{U^{OH}}$-tweezers were optimized by native PAGE analysis. The introduction of mismatches into the arm duplexes enhanced the Gd^{III}-dependent opening of $\mathbf{U^{OH}}$-tweezers, and Design 3 with two mismatches was selected as the optimal sequence. Fluorescence measurement was further conducted to see the effects of the Mg^{II}

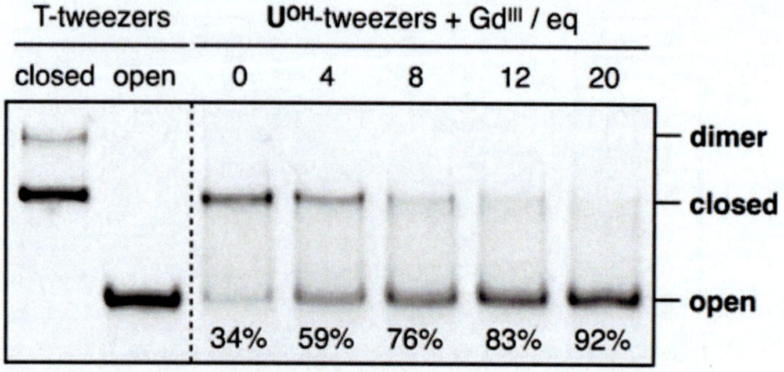

Fig. 4.17 Native PAGE analysis of $\mathbf{U^{OH}}$-tweezers in the presence of varying equivalents of Gd^{III} ions. [each DNA] = 2.0 μM, $[Gd^{III}]$/[each DNA] = 0, 4, 8, 12, and 20 in 10 mM HEPES–NaOH buffer (pH 8.0) containing 100 mM NaCl. 15% gel at 20 °C

concentration and Gd^{III} equivalents on the structural conversion of $\mathbf{U^{OH}}$-tweezers. As a result, 3 mM Mg^{II} ions and 12 equiv. of Gd^{III} ions were selected as the optimal conditions. Since $\mathbf{U^{OH}}$-tweezers showed efficient structural conversion under the optimized conditions, they were expected to be further applicable to Gd^{III}-responsive operation under isothermal conditions.

4.4 Metal-Responsive Operation of DNA Tweezers Containing 5-Hydroxyuracil Bases under Isothermal Conditions

As described in Sect. 4.3, while $\mathbf{U^{OH}}$-tweezers are mainly closed in the absence of Gd^{III} ions, the open form became predominant by the addition of Gd^{III} ions. However, the Gd^{III}-dependent structural conversion was analyzed after annealing the solution in the absence and presence of Gd^{III}. Therefore, no information about the efficiency and kinetics of the Gd^{III}-responsive operation of the tweezers under isothermal conditions had been obtained. In this section, I examined Gd^{III}-responsive reversible operation of $\mathbf{U^{OH}}$-tweezers at a constant temperature based on the conditions optimized in Sect. 4.3.

The Gd^{III}-triggered opening motion of $\mathbf{U^{OH}}$-tweezers was first examined (Fig. 4.18a). A mixture of strands **a3**, $\mathbf{b_{FD}}$, **c3**, and **d** was annealed to form the closed state of the tweezers. The opening operation was initiated by the addition of 12 equiv. of Gd^{III} ions, and fluorescence intensity was recorded in constant time intervals. As shown in Fig. 4.18b, the fluorescence from the FAM was immediately recovered after the addition of Gd^{III} ions. This result suggests that strand **d** was released and the tweezers turned to the open state. The opening operation of $\mathbf{U^{OH}}$-tweezers was nearly completed within 10 min, with a half-completion time ($t_{1/2}$) of less than 0.5 min. The rate of operation was much faster than the Gd^{III}-triggered strand displacement reactions (SDRs) discussed in Chap. 2, which was also driven by the base pair switching of **U^{OH}**. This difference can be explained based on the mechanism of each process (Fig. 4.19). The Gd^{III}-triggered SDRs involved a branch migration process as a rate-limiting step after the interstrand **U^{OH}**–Gd^{III}–**U^{OH}** complexation. In the case of **U^{OH}**-tweezers, strand **d** can no longer bind to the tweezer structures after the addition of Gd^{III} ions due to the destabilization of the **U^{OH}**–A pairs and the intrastrand **U^{OH}**–Gd^{III}–**U^{OH}** complexation. This immediate release of strand **d** will result in a rapid structural conversion to the open form.

Next, the regeneration of the closed tweezers by removing Gd^{III} ions was examined. It was expected that strand **d** can bind to the tweezers again to give the closed structures when Gd^{III} ions are scavenged by chelating agents. As the first trial,

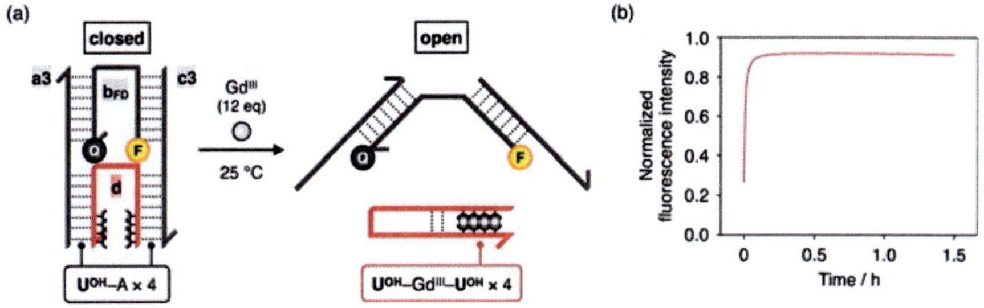

Fig. 4.18 Gd^{III}-responsive opening of **U^{OH}**-tweezers under isothermal conditions. **a** Scheme. F: FAM, Q: Dabcyl. **b** Time-course fluorescence analysis. [each DNA] = 2.0 μM, [Gd^{III}]/[each DNA] = 12 in 10 mM HEPES–NaOH buffer (pH 8.0) containing 100 mM NaCl and 3 mM $MgCl_2$. λ_{ex} = 495 nm, λ_{em} = 519 nm, 25 °C. The recorded fluorescence intensity was normalized by that of the open and closed states of T-tweezers

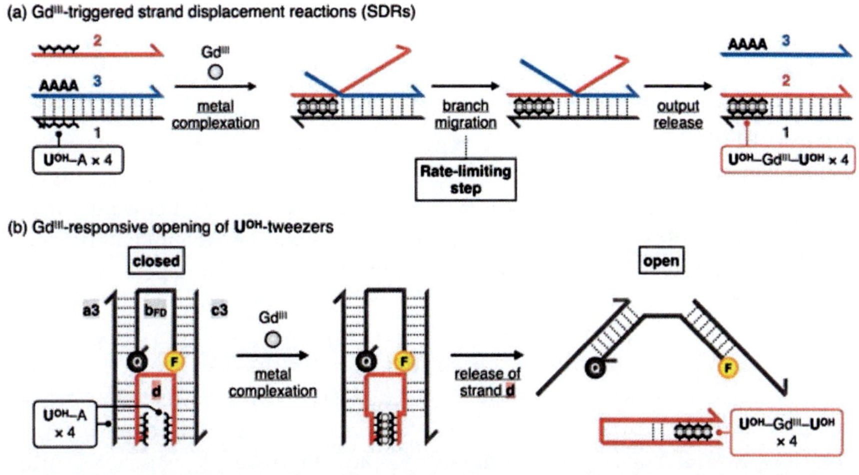

Fig. 4.19 Comparison of the proposed mechanisms of **a** Gd^{III}-triggered strand displacement reactions (SDRs) using **U^{OH}**-containing DNA strands and **b** Gd^{III}-responsive opening of **U^{OH}**-tweezers

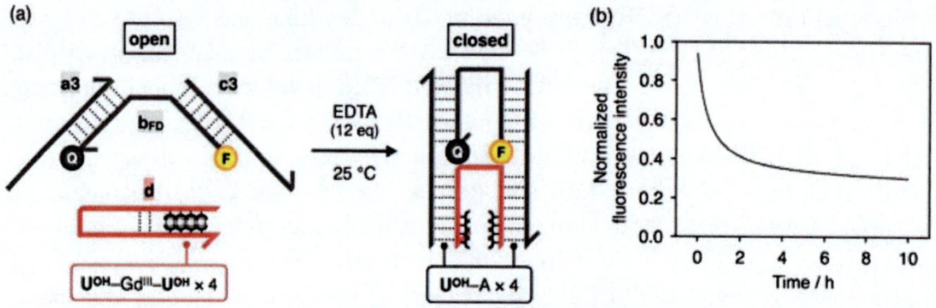

Fig. 4.20 EDTA-responsive closing of **U^{OH}**-tweezers under isothermal conditions. **a** Scheme. F: FAM, Q: Dabcyl. **b** Time-course fluorescence analysis. [each DNA] = 2.0 μM, [Gd^{III}]/[each DNA] = 12, [EDTA]/[each DNA] = 12 in 10 mM HEPES–NaOH buffer (pH 8.0) containing 100 mM NaCl and 3 mM $MgCl_2$. λ_{ex} = 495 nm, λ_{em} = 519 nm, 25 °C. The recorded fluorescence intensity was normalized by that of the open and closed states of T-tweezers

12 equiv. of EDTA were added to the open state of **U^{OH}**-tweezers containing 12 equiv. of Gd^{III} ions (Fig. 4.20a). The fluorescence measurement at 25 °C showed gradual decrease in the emission intensity, indicating the formation of the closed tweezers (Fig. 4.20b). However, the operation speed of the EDTA-triggered closing ($t_{1/2}$ ~ 30 min) was much slower than the Gd^{III}-triggered opening ($t_{1/2}$ < 0.5 min). This slow rate may be due to the Mg^{II} ions in the solution competing with Gd^{III} in complexation with EDTA. Although EDTA shows much higher affinity toward Gd^{III} ions (logK_{ML} = 17.0, 393 K, μ = 0.1, K_{ML} = [EDTA·Gd^{III}]/([EDTA]·[Gd^{III}])) [22] than Mg^{II} (logK_{ML} = 8.82, 393 K, μ = 0.2, K_{ML} = [EDTA·Mg^{II}]/([EDTA]·[Mg^{II}])) [23], this result encouraged me to find more suitable chelating agents to accelerate the closing motion.

For the acceleration of the closing motion, EGTA was used instead of EDTA as a chelating agent (Fig. 4.21a). EGTA shows lower binding affinity to Mg^{II} ions (logK_{ML} = 5.4, 398 K, μ = 0.1, K_{ML} = [EGTA·Mg^{II}]/([EGTA]·[Mg^{II}])) [24] compared with EDTA while it still maintains high coordination strength toward Gd^{III} ions (logK_{ML} = 16.94, 393 K, μ = 0.1, K_{ML} = [EGTA·Gd^{III}]/([EGTA]·[Gd^{III}])) [25] (Fig. 4.21b). Therefore, EGTA was expected to scavenge Gd^{III} ions more efficiently even at high concentrations of Mg^{II} ions. As shown in Fig. 4.21c, the fluorescence from the FAM was gradually quenched by the addition of EGTA, indicating the closing motion of the tweezers. The rate of the fluorescence change ($t_{1/2}$ ~ 4 min) was significantly faster than the EDTA-triggered operation ($t_{1/2}$ ~ 30 min) because of the suppressed interaction between EGTA and Mg^{II}. Overall, the reversible operation of **U^{OH}**-tweezers was demonstrated using Gd^{III} ions and EGTA as the inputs.

Figure 4.22 shows a proposed mechanism for the EGTA-triggered closing of **U^{OH}**-tweezers. After the removal of Gd^{III} ions with EGTA, the resultant Gd^{III}-free strand **d** binds to the tweezers to yield the closed form. Here, the hybridization rate of strand **d** with the tweezer structures was investigated. The closing operation was conducted just by the addition of strand **d** to the open tweezers (Fig. 4.23a). The fluorescence from

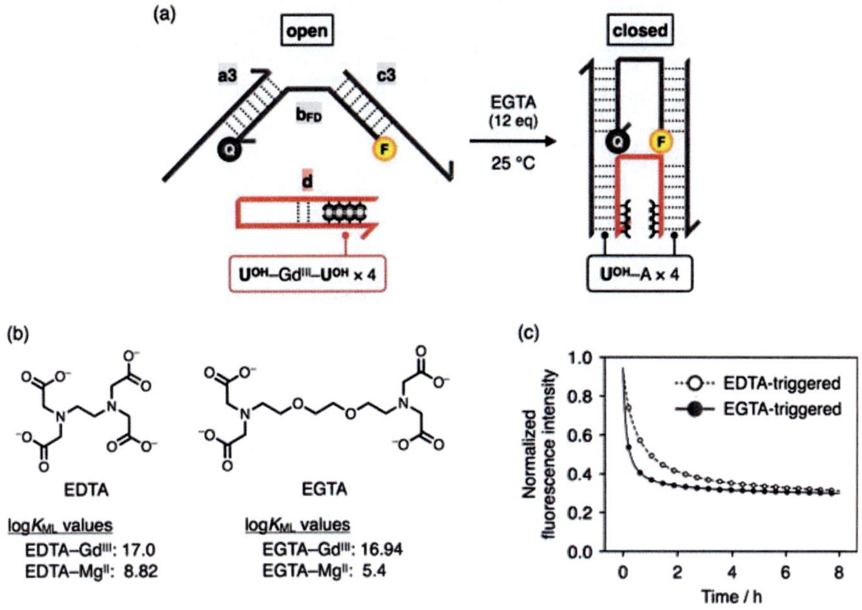

Fig. 4.21 EGTA-responsive closing of $\mathbf{U^{OH}}$-tweezers under isothermal conditions. **a** Scheme. F: FAM, Q: Dabcyl. **b** Chemical structures and logK_{ML} values of EDTA and EGTA. **c** Time-course fluorescence analysis compared with the EDTA-triggered operation. [each DNA] = 2.0 μM, [Gd^{III}]/[each DNA] = 12, [EDTA or EGTA]/[each DNA] = 12 in 10 mM HEPES–NaOH buffer (pH 8.0) containing 100 mM NaCl and 3 mM $MgCl_2$. λ_{ex} = 495 nm, λ_{em} = 519 nm, 25 °C. The recorded fluorescence intensity was normalized by that of the open and closed states of T-tweezers

the FAM was immediately quenched by the addition of strand **d** ($t_{1/2}$ < 0.5 min) (Fig. 4.23b), exhibiting a faster change than the EGTA-triggered closing ($t_{1/2}$ ~ 4 min). According to this result, it was concluded that the binding of Gd^{III}-free strand **d** to the tweezers is much faster than the whole process of the EGTA-triggered closing. Thus, the mechanism shown in Fig. 4.22 is supported and the rate-limiting step of the closing operation is the removal of Gd^{III} ions from strand **d**.

Finally, iterative switching of $\mathbf{U^{OH}}$-tweezers between the closed and open forms was conducted by the repeated addition of Gd^{III} ions and EGTA (Fig. 4.24a). To a preannealed mixture of strands **a3**, $\mathbf{b_{FD}}$, **c3**, and **d** forming the closed tweezers,

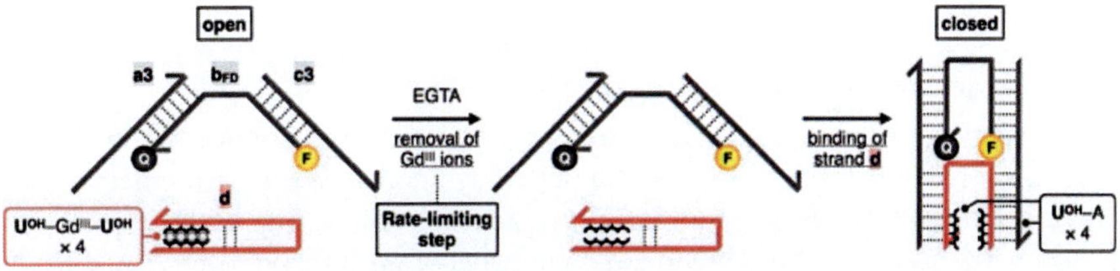

Fig. 4.22 The proposed mechanism of EGTA-triggered closing of $\mathbf{U^{OH}}$-tweezers

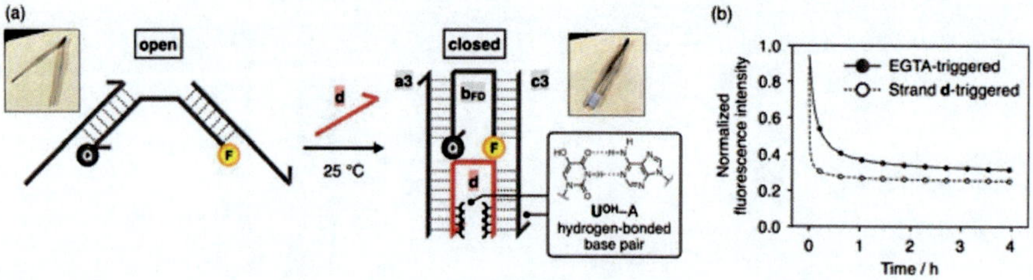

Fig. 4.23 Closing of **U^OH**-tweezers triggered by the addition of strand **d**. **a** Scheme. F: FAM, Q: Dabcyl. **b** Time-course fluorescence analysis compared with the EGTA-triggered operation. [each DNA] = 2.0 μM, [Gd^{III}]/[each DNA] = 0 or 12, [EGTA]/[each DNA] = 0 or 12 in 10 mM HEPES–NaOH buffer (pH 8.0) containing 100 mM NaCl and 3 mM $MgCl_2$. $\lambda_{ex} = 495$ nm, $\lambda_{em} = 519$ nm, 25 °C. The recorded fluorescence intensity was normalized by that of the open and closed states of T-tweezers

12 equiv. of Gd^{III} ions and EGTA were sequentially added at 25 °C at a time interval of 2 h. Native PAGE analysis indicated that the open and closed forms of $\mathbf{U^{OH}}$-tweezers were alternately yielded in response to Gd^{III} ions and EGTA (Fig. 4.24b). Furthermore, time-course fluorescence measurement revealed that the emission from the FAM was repeatedly increased and decreased for four cycles by the sequential addition of Gd^{III} ions and EGTA (Fig. 4.24c). Although a slight decrease in the switching efficiency was observed, this result confirmed the reversible opening and closing of $\mathbf{U^{OH}}$-tweezers in response to Gd^{III} ions.

In summary, it was demonstrated that $\mathbf{U^{OH}}$-tweezers can be reversibly operated in response to Gd^{III} ions under isothermal conditions. By utilizing the base pair switching of $\mathbf{U^{OH}}$ from the $\mathbf{U^{OH}}$–A pairs to the intrastrand $\mathbf{U^{OH}}$–Gd^{III}–$\mathbf{U^{OH}}$ pairs, the structural conversion of the DNA tweezers smoothly proceeded. The base pair switching of $\mathbf{U^{OH}}$ would be further applicable to the control of higher-ordered DNA nanoarchitectures and dynamic nanodevices with more advanced functions.

4.5 Discussion

I have developed Gd^{III}-responsive DNA tweezers containing $\mathbf{U^{OH}}$ bases, named as $\mathbf{U^{OH}}$-tweezers. The $\mathbf{U^{OH}}$ bases were introduced into the tweezers based on the base sequences of the original DNA tweezers [11]. The formation of $\mathbf{U^{OH}}$-tweezers was confirmed by native PAGE analysis using tweezers with natural thymine (T) instead of $\mathbf{U^{OH}}$ (T-tweezers) as a control. The base sequences of the tweezer arms were optimized for the Gd^{III}-dependent structural conversion. As a result, the DNA tweezers containing two mismatch pairs in each arm were selected. Furthermore, the structural conversion of $\mathbf{U^{OH}}$-tweezers was also supported by fluorescence measurement of the tweezers with a FAM–Dabcyl pair. Based on the results of the fluorescence analysis, the optimal Mg^{II} concentration and Gd^{III} equivalents were found to be 3 mM Mg^{II} and 12 equiv. of Gd^{III} ions, respectively.

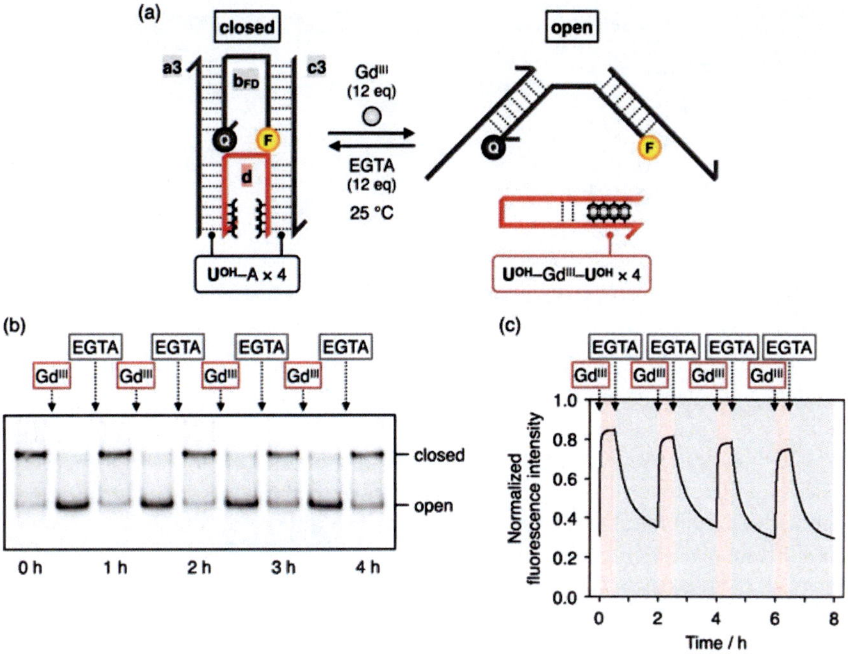

Fig. 4.24 Repeated opening and closing operations of $\mathbf{U^{OH}}$-tweezers by the alternate addition of Gd^{III} ions and EGTA. **a** Scheme. F: FAM, Q: Dabcyl. **b** Native PAGE analysis. Strand $\mathbf{b_F}$ without a Dabcyl was used instead of $\mathbf{b_{FD}}$. **c** Time-course fluorescence measurement. [each DNA] = 2.0 μM, [Gd^{III}]/[each DNA] = 12, [EGTA]/[each DNA] = 12 in 10 mM HEPES–NaOH buffer (pH 8.0) containing 100 mM NaCl and 3 mM $MgCl_2$. Native PAGE: 15% gel at 20 °C. Fluorescence measurement: λ_{ex} = 495 nm, λ_{em} = 519 nm, 25 °C. The recorded fluorescence intensity was normalized by that of the open and closed states of T-tweezers

Subsequently, Gd^{III}-responsive operation of $\mathbf{U^{OH}}$-tweezers was conducted at a constant temperature with the optimized conditions (Fig. 4.25). The opening motion of $\mathbf{U^{OH}}$-tweezers smoothly proceeded by the addition of 12 equiv. of Gd^{III} ions to the closed tweezers. This structural conversion was probably driven by the destabilization of the $\mathbf{U^{OH}}$–A pairs and the intrastrand $\mathbf{U^{OH}}$–Gd^{III}–$\mathbf{U^{OH}}$ complexation in $\mathbf{U^{OH}}$-containing strand **d**. The closing operation was also demonstrated by removing Gd^{III} ions with chelating agents. Although the EDTA-triggered closing was much slower than the Gd^{III}-responsive opening, the operation rate was improved using EGTA instead. Finally, repeated operation of $\mathbf{U^{OH}}$-tweezers between the closed and open forms was demonstrated by the sequential addition of Gd^{III} ions and EGTA as the inputs.

The structural conversion of $\mathbf{U^{OH}}$-tweezers was triggered by the intrastrand metal complexation of closing strand **d**. This working principle is unique compared with the previously reported DNA tweezers that are operated by toehold-mediated

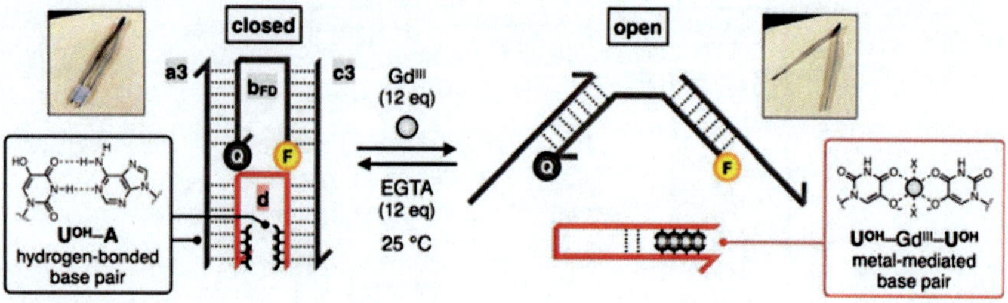

Fig. 4.25 Gd^{III}-responsive reversible control of **U^{OH}**-tweezers driven by the base pair switching of **U^{OH}**

strand displacement reactions (SDRs). Since the **U^{OH}**-containing strand was immediately dissociated via the destabilization of **U^{OH}**–A pairs and the intrastrand **U^{OH}**–Gd^{III}–**U^{OH}** complexation, the Gd^{III}-responsive opening of **U^{OH}**-tweezers smoothly proceeded. The rate of the Gd^{III}-triggered opening motion ($t_{1/2} < 0.5$ min) was comparable with the reported DNA tweezers working in a time scale of minutes. In addition, inherent reversibility of metal complexes also played a key role in the reversible control of **U^{OH}**-tweezers. In fact, the closed tweezers were restored to the same level as the Gd^{III}-free conditions by the removal of Gd^{III} ions with EGTA. It should also be noted that the tweezer structures were reversibly switched without significant decrease in the efficiency of the structural conversion.

DNA tweezers have been utilized as a regulator of chemical and biological functions. For example, catalytic activity of enzymes can be switched by modifying the termini of the tweezer arms with a pair of an enzyme and a cofactor [14]. Furthermore, it is also possible to capture target proteins via aptamer sequences introduced into the termini of the tweezers [13]. In this context, **U^{OH}**-tweezers developed in this study are expected to be applicable to functional switching. Since **U^{OH}** bases were introduced only in the closing strand, the open form of the tweezers consists of purely natural sequences. Therefore, **U^{OH}**-tweezers are highly compatible with additional modification such as elongation of the arms with functional sequences and chemical modification with enzymes or other small molecules. It is expected that **U^{OH}**-tweezers would be applicable as a novel motif of metal-responsive functional regulators.

This study demonstrates that the DNA tweezers can be operated by the base pair switching of **U^{OH}**. Considering that DNA tweezers are one of the simplest types of DNA-based molecular machines, the base pair switching of **U^{OH}** would be further applicable to more complicated DNA-based systems. **U^{OH}**-tweezers were operated by regulating the binding of closing strand **d** to the two tweezer arms. This strategy, which uses **U^{OH}**-containing strands as a reversibly-binding effector, could be applied to the control of higher-ordered DNA nanostructures such as DNA tiles and origami. For example, assembly of DNA tiles can be controlled in a Gd^{III}-responsive manner by connecting multiple tile structures with **U^{OH}**-containing strands. Furthermore, Gd^{III}-dependent dynamic structural conversion of DNA origami structures would be possible by introducing **U^{OH}** bases into the staple strands. Because the **U^{OH}**

bases form hydrogen-bonded **U^{OH}**–A pairs in a manner similar to natural T–A pairs, any natural sequences in DNA tiles and origami can be readily modified with **U^{OH}** bases. Since various DNA nanodevices composed of DNA tiles or origami have been actively studied, there is a great chance to develop advanced DNA molecular machines driven by the metal-mediated base pair switching of **U^{OH}**.

4.6 Experimental Section

General

All natural DNA strands including FAM-modified strands and Dabcyl-modified strands were purchased from Japan Bio Service Co., Ltd. (Saitama, Japan) at HPLC purification grade. DNA strands containing 5-hydroxyuracil ($\mathbf{U^{OH}}$) nucleobases were synthesized as follows. DNA concentrations were calculated based on UV absorbance at 260 nm. Metal sources were purchased from Soekawa Chemical Co. ($GdCl_3{\cdot}6H_2O$ (99.9%)). Fluorescence spectra were recorded on a JASCO FP-8300 fluorometer with a path length of 0.5 cm. Native polyacrylamide gel electrophoresis was carried out with 15% polyacrylamide (acrylamide: *N*,*N*′-methylenebis(acrylamide) = 19: 1). The gels were analyzed using Gel Doc EZ Imager and Image Lab software (Bio-Rad).

DNA Synthesis

DNA strands containing 5-hydroxyuracil ($\mathbf{U^{OH}}$) nucleobases were synthesized on an Applied Biosystems 394 DNA synthesizer on a 1-μmol scale in a DMTr-on mode with ultramild deprotection phosphoramidites and reagents (Glen Research). DNA synthesis was conducted following the standard procedure except for an extended coupling time for the $\mathbf{U^{OH}}$ nucleoside (15 min). The synthesized strands were cleaved from the solid support and deprotected using a 25% NH_3 aqueous solution (1.0 mL) at room temperature for 4 h. The obtained DNA strands were first detritylated and purified by a Poly-Pak II cartridge (Glen Research). After further purification by denaturing PAGE, impurities were removed by size-exclusion chromatography and isopropanol precipitation. Finally, all DNA strands were identified by matrix assisted laser desorption ionization-time-of-flight (MALDI-TOF) mass spectrometry. The amount of the DNA strands was determined based on the UV absorbance at 260 nm. The molar extinction coefficients (ε_{260}) of the strands were calculated based on the nearest-neighbor method ($\varepsilon_{260} = 4.59 \times 10^3$ M^{-1} cm^{-1} for the $\mathbf{U^{OH}}$ nucleoside). FAM- or Dabcyl-labeled strands were quantified based on UV absorbance at 495 nm and 476 nm ($\varepsilon_{495} = 7.50 \times 10^3$ M^{-1} cm^{-1} for FAM, $\varepsilon_{476} = 3.20 \times 10^3$ M^{-1} cm^{-1} for Dabcyl), respectively. Strand $\mathbf{b_{FD}}$ with both the FAM and Dabcyl was quantified based on UV absorbance at 260 nm ($\varepsilon_{260} = 2.09 \times 10^4$ M^{-1} cm^{-1} for FAM, $\varepsilon_{260} = 7.28 \times 10^3$ M^{-1} cm^{-1} for Dabcyl).

Strand **d**. 5′ - G**U^OH^U^OH^** **U^OH^U^OH^**G GTG TTA GTC ATG CCA GTG TCA CAC G**U^OH^U^OH^** **U^OH^U^OH^**G - 3′. MALDI MS: *m*/*z* calcd for $[C_{347}H_{432}N_{128}O_{230}P_{35} - H]^-$: 11,158.0, found: 11,158.1.

MALDI-TOF Mass Spectrometry of DNA Strands

The sample was desalted by size-exclusion chromatography (Sephadex G-25 Fine). After treated with a cation-exchange resin (DOWEX 50Wx8, NH_4^+ form) to remove remaining sodium ions, the samples were mixed with 3-hydroxypicolinic acid (3-HPA) and ammonium citrate as a matrix. Oligonucleotide Calibration Standard (Bruker) was used as a calibration standard. MALDI-TOF mass spectra were recorded on Bruker ultrafleXtreme and Autoflex III in a negative mode.

Simulation of Melting Profiles of DNA Tweezers by NUPACK Software

Melting profiles of the duplex arms of $\mathbf{U^{OH}}$-tweezers were simulated by NUPACK software [20, 21]. To simulate the hybridization under conditions in the absence and presence of Gd^{III} ions, the $\mathbf{U^{OH}}$–A pairs were replaced by T–A pairs and T–T mismatches, respectively. The melting profiles were calculated from 4 °C to 85 °C with the temperature increment of 3 °C. Three-stranded assemblies were included in the calculation as a maximum complex size. Concentrations of each DNA strand and Na^I ions were set at 2 μM and 0.1 M, respectively. In the simulated melting profile, fractions of the nucleobases which do not form stable base pairs were plotted against temperature.

Formation of Control DNA Tweezers Containing T Bases (T-Tweezers)

The samples were prepared by mixing strands **a1**, $\mathbf{b_F}$, **c1**, and $\mathbf{d_T}$ (2.0 μM each) in 10 mM HEPES–NaOH buffer (pH 8.0) containing 100 mM NaCl. The prepared solutions were annealed (85 °C → 25 °C, − 1.0 °C/min) and analyzed by native PAGE with 15% gel at 25 °C. The bands were detected by the fluorescence of the FAM on strand $\mathbf{b_F}$.

Optimization of the Base Sequences of the $\mathbf{U^{OH}}$-Containing DNA Tweezers ($\mathbf{U^{OH}}$-Tweezers) by Native PAGE Analysis

The samples were prepared by mixing strands **a**, $\mathbf{b_F}$, **c**, and **d** (2.0 μM each) in 10 mM HEPES–NaOH buffer (pH 8.0) containing 100 mM NaCl. For strands **a** and **c**, three different combinations (**a1** and **c1**, **a2** and **c2**, and **a3** and **c3**) were used. After annealing the prepared solutions (85 °C → 25 °C, − 1.0 °C/min) in the absence and presence of Gd^{III} ions (4 equiv.), native PAGE analysis was performed in TAMg buffer (40 mM Tris, 7.6 mM $MgCl_2$, pH 8.0 adjusted by AcOH) with 15% gel at 25 °C. The closed and open forms of T-tweezers were also subjected to the analysis as authentic samples. The bands were detected by the fluorescence of the FAM on strand $\mathbf{b_F}$.

Fluorescence Measurement of the DNA Tweezers

The samples were prepared by mixing strands **a3**, $\mathbf{b_{FD}}$, **c3**, and **d** or $\mathbf{d_T}$ (2.0 μM each) in 10 mM HEPES–NaOH buffer (pH 8.0) containing 100 mM NaCl. The prepared solutions were annealed (85 °C → 25 °C, – 1.0 °C/min) in the absence and presence of Gd^{III} ions (4 equiv.) The fluorescence spectra were measured at 25 °C ($\lambda_{ex} = 495$ nm). The yield of the open tweezers was calculated by normalizing the observed fluorescence intensity (I) with the intensity of the closed and open states of T-tweezers (I_{closed} and I_{open}, respectively) as follows:

$$\text{Yield of the open tweezers}(\%) = \left\{(I - I_{closed})/\left(I_{open} - I_{closed}\right)\right\} \cdot 100.$$

Optimization of Mg^{II} Concentration by Fluorescence Measurement of the DNA Tweezers

The samples were prepared by mixing strands **a3**, $\mathbf{b_{FD}}$, **c3**, and **d** or $\mathbf{d_T}$ (2.0 μM each) in 10 mM HEPES–NaOH buffer (pH 8.0) containing 100 mM NaCl. The prepared solutions were annealed (85 °C → 25 °C, – 1.0 °C/min) in the absence and presence of Gd^{III} ions (4 equiv.) After the fluorescence spectra were recorded, Mg^{II} aqueous solution was titrated to the samples and the measurement was repeatedly conducted with variable Mg^{II} concentrations. After every addition of Mg^{II} ions, the solution was incubated at 25 °C for 30 min before the measurement. The fluorescence spectra were measured at 25 °C ($\lambda_{ex} = 495$ nm). The yield of the open tweezers at each Mg^{II} concentration was calculated by normalizing the observed fluorescence intensity (I) with the intensity of the closed and open states of T-tweezers (I_{closed} and I_{open}, respectively) as follows:

$$\text{Yield of the open tweezers}(\%) = \left\{(I - I_{closed})/\left(I_{open} - I_{closed}\right)\right\} \cdot 100.$$

Optimization of Gd^{III} Equivalents for Structural Conversion of U^{OH}-Tweezers by Fluorescence Measurement

The samples were prepared by mixing strands **a3**, $\mathbf{b_{FD}}$, **c3**, and **d** (2.0 μM each) in 10 mM HEPES–NaOH buffer (pH 8.0) containing 100 mM NaCl and 2 or 3 mM $MgCl_2$. After annealing the sample (85 °C → 25 °C, – 1.0 °C/min) in the presence of Gd^{III} ions (4 equiv.), the fluorescence spectra were recorded. Then Gd^{III} aqueous solution was titrated to the samples and the measurement was repeatedly conducted with variable Gd^{III} concentrations. After every addition of Gd^{III}, the solution was incubated at 25 °C for 30 min before the measurement. The fluorescence spectra were measured at 25 °C ($\lambda_{ex} = 495$ nm). The yield of the open tweezers was calculated as described above.

Optimization of Gd^{III} Equivalents for Structural Conversion of U^{OH}-Tweezers by Native PAGE

The samples were prepared by mixing strands **a3**, $\mathbf{b_F}$, **c3**, and **d** (2.0 μM each) in 10 mM HEPES–NaOH buffer (pH 8.0) containing 100 mM NaCl, 3 mM $MgCl_2$. The prepared solutions were annealed (85 °C → 25 °C, – 1.0 °C/min) in the absence and presence of Gd^{III} ions (4, 8, 12, and 20 equiv.) and subjected to native PAGE analysis in TAMg buffer (40 mM Tris, 3 mM $MgCl_2$, pH 8.0 adjusted by AcOH) with 15% gel at 20 °C. The closed and open forms of T-tweezers were also analyzed as authentic samples. The bands were detected by the fluorescence of the FAM on strand $\mathbf{b_F}$. The yields of the closed and open forms were estimated by comparing the intensities of the two bands. The average of the three independent experiments was summarized as the yield of each structure.

Time-Course Fluorescence Analysis of Gd^{III}-Triggered Opening of U^{OH}-Tweezers

The samples were prepared by mixing strands **a3**, $\mathbf{b_{FD}}$, **c3**, and **d** (2.0 μM each) in 10 mM HEPES–NaOH buffer (pH 8.0) containing 100 mM NaCl and 3 mM $MgCl_2$. After annealing the sample (85 °C → 25 °C, − 1.0 °C/min), Gd^{III} ions (12 equiv.) were added and the fluorescence intensity was recorded at 25 °C in 30-s intervals ($\lambda_{ex} = 495$ nm, $\lambda_{em} = 519$ nm). Normalized fluorescence intensity was calculated by normalizing the observed fluorescence intensity (I) with the intensity of the closed and open states of T-tweezers (I_{closed} and I_{open}, respectively) as follows:

$$\text{Normalized fluorescence intensity} = (I - I_{closed})/(I_{open} - I_{closed}).$$

Time-Course Fluorescence Analysis of U^{OH}-Tweezers Closing by Chelating Agents

The samples were prepared by mixing strands **a3**, $\mathbf{b_{FD}}$, **c3**, and **d** (2.0 μM each) in 10 mM HEPES–NaOH buffer (pH 8.0) containing 100 mM NaCl and 3 mM $MgCl_2$. After the addition of Gd^{III} ions (12 equiv.), the samples were annealed (85 °C → 25 °C, − 1.0 °C/min). Then EDTA or EGTA (12 equiv.) was added to initiate the structural change. The fluorescence intensity was recorded at 25 °C in 30-s intervals ($\lambda_{ex} = 495$ nm, $\lambda_{em} = 519$ nm). In the control experiment, strand **d** was added to a preannealed mixture of strands **a3**, $\mathbf{b_{FD}}$, and **c3**. The fluorescence intensity was normalized as described above.

Repeated Operation of U^{OH}-Tweezers between the Closed and Open Forms

For native PAGE analysis, the samples were prepared by mixing strands **a3**, $\mathbf{b_F}$, **c3**, and **d** (2.0 μM each) in 10 mM HEPES–NaOH buffer (pH 8.0) containing 100 mM NaCl and 3 mM $MgCl_2$. After annealing the sample (85 °C → 25 °C, − 1.0 °C/min), Gd^{III} ions (12 equiv.) and EGTA (12 equiv.) were sequentially added and each sample was analyzed by native PAGE in TAMg buffer (40 mM Tris, 3 mM $MgCl_2$, pH 8.0 adjusted by AcOH) with 15% gel at 20 °C.

For time-course fluorescence analysis, the samples were prepared by mixing strands **a3**, $\mathbf{b_{FD}}$, **c3**, and **d** (2 μM each) in 10 mM HEPES–NaOH buffer (pH 8.0) containing 100 mM NaCl and 3 mM $MgCl_2$. After annealing the sample (85 °C → 25 °C, − 1.0 °C/min), Gd^{III} ions (12 equiv.) and EGTA (12 equiv.) were sequentially added and the fluorescence intensity was recorded at 25 °C in 30-s intervals ($\lambda_{ex} = 495$ nm, $\lambda_{em} = 519$ nm). The fluorescence intensity was normalized as described above.

References

1. Seeman NC, Sleiman HF (2017) Nat Rev Mater 3:17068
2. Seeman NC (2016) Structural DNA nanotechnology. Cambridge University Press, Cambridge
3. Zhang F, Nangreave J, Liu Y, Yan H (2014) J Am Chem Soc 136:11198–11211
4. Hu Q, Li H, Wang L, Gu H, Fan C (2019) Chem Rev 119:6459–6506
5. Xiao M, Lai W, Man T, Chang B, Li L, Chandrasekaran AR, Pei H (2019) Chem Rev 119:11631–11717
6. Zhang DY, Seelig G (2011) Nat Chem 3:103–113
7. Simmel FC, Yurke B, Singh HR (2019) Chem Rev 119:6326–6369
8. Mao X, Liu M, Li Q, Fan C, Zuo X (2022) JACS Au 2:2381–2399
9. Guo Y, Wei B, Xiao S, Yao D, Li H, Xu H, Song T, Li X, Liang H (2017) Quant Biol 5:25–41
10. He S, Ge Z, Zuo X, Fan C, Mao X (2021) NPG Asia Mater 13:42
11. Yurke B, Turberfield AJ, Mills AP Jr, Simmel FC, Neumann JL (2000) Nature 406:605–608
12. Fu T-J, Seeman NC (1993) Biochemistry 32:3211–3220
13. Zhou C, Yang Z, Liu D (2012) J Am Chem Soc 134:1416–1418
14. Liu M, Fu J, Hejesen C, Yang Y, Woodbury NW, Gothelf K, Liu Y, Yan H (2013) Nat Commun 4:2127
15. Lu S, Shen J, Fan C, Li Q, Yang X (2021) Adv Sci 8:2100328
16. Liang X, Nishioka H, Takenaka N, Asanuma H (2008) ChemBioChem 9:702–705
17. Kamiya Y, Asanuma H (2014) Acc Chem Res 47:1663–1672
18. Elbaz J, Wang Z-G, Orbach R, Willner I (2009) Nano Lett 9:4510–4514
19. Assi HA, Garavís M, González C, Damha MJ (2018) Nucleic Acids Res 46:8038–8056
20. Zadeh JN, Steenberg CD, Bois JS, Wolfe BR, Pierce MB, Khan AR, Dirks RM, Pierce NA (2011) J Comput Chem 32:170–173
21. Fornace ME, Huang J, Newman CT, Porubsky NJ, Pierce MB, Pierce NA (2022). ChemRxiv. https://doi.org/10.26434/chemrxiv-2022-xv98l
22. Wheelwright EJ, Spedding FH, Schwarzenbach G (1953) J Am Chem Soc 75:4196–4201
23. Arena G, Musumeci S, Purrello R (1983) Thermochim Acta 61:129–138
24. Schmid RW, Reiley CN (1957) Anal Chem 29:264–268
25. Mackey JL, Hiler MA, Powell JE (1962) J Phys Chem 66:311–314

Chapter 5
Conclusion

Abstract In this study, coordination-driven base pair switching of modified uracil bases was proposed and applied to dynamic control of DNA structures. DNA strand displacement reactions essential for DNA structural conversion were successfully controlled in a Gd^{III}-responsive manner by introducing 5-hydroxyuracil (**U^{OH}**) bases at the termini of DNA strands. To resolve the sequence limitation of the **U^{OH}** bases, an *N*,*N*-dicarboxymethyl-5-aminouracil (**dcaU**) base with an iminodiacetate ligand was developed and the Gd^{III}-dependent base pair switching was achieved with a single incorporation of **dcaU**. Finally, it was demonstrated that **U^{OH}**-modified DNA tweezers were reversibly opened and closed upon addition and removal of Gd^{III} ions, respectively, under isothermal conditions. The introduction of modified uracil bases instead of natural T bases has little effect on the thermal stability and structures of DNA assemblies. Therefore, metal-responsive base pair switching is expected to be widely applicable to the dynamic control of DNA supramolecules. This chapter summarizes the achievements of this study and overviews future perspectives of the "base pair switching" technology.

Keywords Base pair switching · Modified uracil · DNA strand displacement reaction · Molecular machine · DNA nanotechnology

5.1 Conclusion

DNA nanotechnology has been recognized as a research field in which various nanostructures can be constructed based on sequence-specific self-assembly of DNA molecules [1–3]. Researchers have achieved dynamic control of the DNA assemblies, featured as "dynamic DNA nanotechnology" [4, 5]. In particular, it has gathered increasing attention to induce structural conversion of DNA in response to external stimuli other than oligonucleotide binding. In this study, I aimed to utilize metal complexation as a trigger for inducing dynamic structural conversion of DNA assemblies. For this purpose, I have demonstrated metal-responsive "base pair switching" of 5-modified uracil (**U^{X}**) bases as a tool for dynamic control of DNA structures.

K. Mori, *Metal-Responsive Base Pair Switching of Ligand-type Uracil Nucleobases*, Springer Theses, https://doi.org/10.1007/978-981-99-9400-7_5

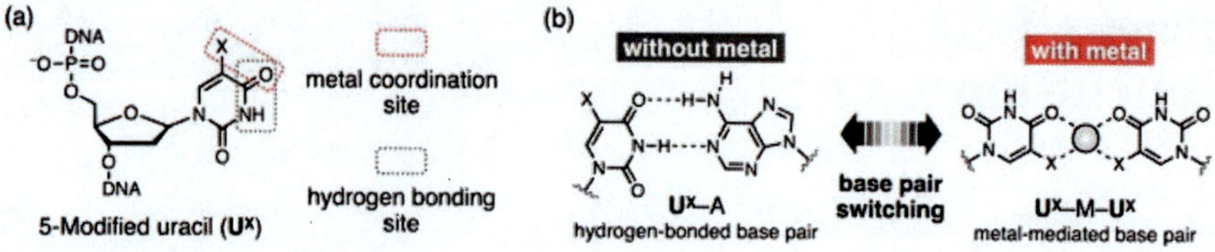

Fig. 5.1 **a** Bifacial structure of a 5-modified uracil (**U**$^{\mathbf{X}}$) base. **b** Metal-dependent base pair switching of the **U**$^{\mathbf{X}}$ base

The **U**$^{\mathbf{X}}$ bases have a metal coordination site composed of the 5-substituent (X) and 4-carbonyl groups while it retains the hydrogen-bonding site of natural uracil bases (Fig. 5.1a). Therefore, the **U**$^{\mathbf{X}}$ bases can form both hydrogen-bonded **U**$^{\mathbf{X}}$–A base pairs and metal-mediated **U**$^{\mathbf{X}}$–M–**U**$^{\mathbf{X}}$ (M = metal) pairs within DNA duplexes (Fig. 5.1b). Based on this "bifacial" base-pairing property, DNA assemblies were dynamically controlled through metal-dependent interconversion of the two base-pairing modes, termed as base pair switching.

Previously, it was revealed that the 5-hydroxyuracil (**U**$^{\mathbf{OH}}$) bases behave as a bidentate ligand through the coordination of the 4-carbonyl and 5-hydroxy groups, resulting in their high affinity toward lanthanide ions [6]. Therefore, the **U**$^{\mathbf{OH}}$ base shows bifacial base-pairing behaviors with the hydrogen-bonded **U**$^{\mathbf{OH}}$–A and metal-mediated **U**$^{\mathbf{OH}}$–Gd^{III}–**U**$^{\mathbf{OH}}$ pairing. In Chap. 2, I demonstrated the Gd^{III}-responsive base pair switching of the **U**$^{\mathbf{OH}}$ base (Fig. 5.2a, b). Duplex melting experiments revealed that a DNA duplex with four **U**$^{\mathbf{OH}}$–**U**$^{\mathbf{OH}}$ pairs at the termini was thermally stabilized by the addition of Gd^{III} ions due to the formation of the **U**$^{\mathbf{OH}}$–Gd^{III}–**U**$^{\mathbf{OH}}$ pairs. In contrast, a duplex with **U**$^{\mathbf{OH}}$–A pairs was destabilized upon addition of Gd^{III} ions. As a result, the relative stability of the two duplexes was inverted, indicating that the base-pairing partner of **U**$^{\mathbf{OH}}$ can be switched by the addition of Gd^{III} ions (Fig. 5.2c).

Furthermore, the Gd^{III}-dependent base pair switching was applied to DNA strand displacement reactions (SDRs) [7], which is an essential tool in dynamic DNA nanotechnology (Fig. 5.2d). In the standard SDRs, an output DNA strand is displaced from an original duplex by invasion of an input strand, resulting in the formation of another duplex with higher thermal stability. The Gd^{III}-triggered SDR was conducted by adding 4 equiv. of Gd^{III} ions to a mixture of original duplex **1**·**3** with **U**$^{\mathbf{OH}}$–A pairs and **U**$^{\mathbf{OH}}$-containing strand **2**. Time-course fluorescence measurement revealed that the SDR proceeded to release output strand **3** and yield duplex **1**·**2** containing **U**$^{\mathbf{OH}}$–Gd^{III}–**U**$^{\mathbf{OH}}$ pairs. By adding 4 equiv. of EDTA to scavenge Gd^{III} ions, the reverse SDR also proceeded to regenerate duplex **1**·**3**.

Most of the previous studies on the metal-dependent control of DNA assemblies relied on the stabilization of only the structures containing input metal ions. In contrast, the 5-modified uracil (**U**$^{\mathbf{X}}$) bases can stabilize two different states through the different base-pairing modes, the hydrogen-bonded **U**$^{\mathbf{X}}$–A and metal-mediated **U**$^{\mathbf{X}}$–M–**U**$^{\mathbf{X}}$ pairs. Therefore, efficient structural conversion between two stable assemblies can be realized using the metal-responsive base pair switching.

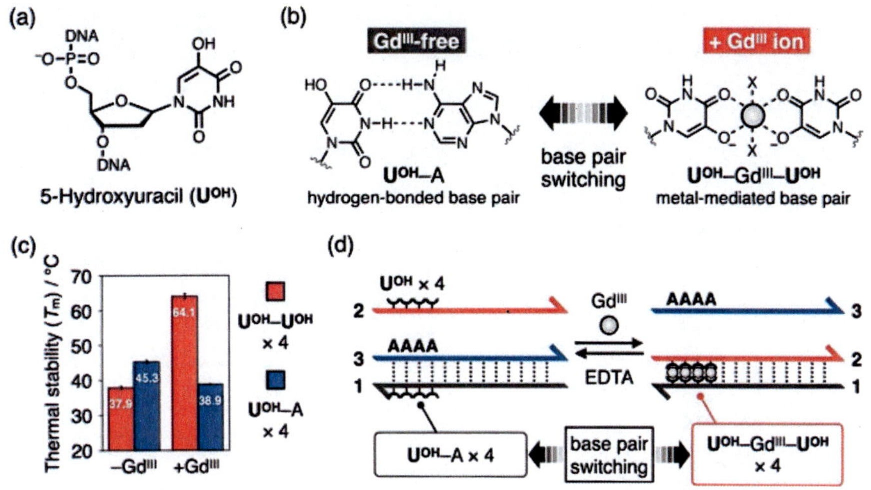

Fig. 5.2 **a** Structure of a 5-hydroxyuracil (**U^OH^**) base. **b** Gd^{III}-dependent base pair switching of the **U^OH^** base. "X" denotes additional ligands coordinating to the Gd^{III} ion such as H_2O and the neighboring nucleobases. **c** Melting temperatures (T_m) of DNA duplexes containing **U^OH^–U^OH^** or **U^OH^**–A pairs in the absence and presence of Gd^{III} ions (4 equiv.). **d** Gd^{III}-triggered DNA strand displacement reactions (SDRs) using **U^OH^**-containing DNA strands

Based on this feature, the SDRs were reversibly controlled without any exposed toehold sequences. Considering broad applicability of the SDRs in dynamic DNA nanotechnology, the metal-triggered SDRs demonstrated in this study would be a general tool to develop metal-responsive DNA supramolecules. Based on the toehold-less design using the $\mathbf{U^X}$ base, the design strategy of stimuli-responsive DNA systems would be broadened.

Although Gd^{III}-responsive base pair switching of $\mathbf{U^{OH}}$ was successfully demonstrated, it has been required to consecutively introduce multiple $\mathbf{U^{OH}}$ bases to form $\mathbf{U^{OH}}$–Gd^{III}–$\mathbf{U^{OH}}$ pairs. In Chap. 3, I aimed to develop a bifacial nucleobase that forms a single metal-mediated base pair inside DNA duplexes. For this purpose, an iminodiacetic acid (IDA) ligand was introduced at the 5 position of the uracil and an *N*,*N*-dicarboxymethyl-5-aminouracil (**dcaU**) base was designed (Fig. 5.3a). The **dcaU** base was expected to show improved coordination strength toward Gd^{III} ions compared with $\mathbf{U^{OH}}$ due to the negative charges and chelating effects of the IDA ligand. Since the **dcaU** base retains the hydrogen-bonding site of natural pyrimidine bases, it can show the base pair switching between hydrogen-bonded **dcaU**–A and metal-mediated **dcaU**–Gd^{III}–**dcaU** pairs (Fig. 5.3b). Metal complexation studies using the **dcaU** nucleoside confirmed that a 2:1 complex of **dcaU** and a Gd^{III} ion can be stably formed. According to duplex melting experiments, duplex **7·8** with a **dcaU–dcaU** pair showed Gd^{III}-dependent thermal stabilization, indicating the formation of a single **dcaU**–Gd^{III}–**dcaU** pair. Duplex **7·8A** with a **dcaU**–A pair was slightly destabilized by the addition of Gd^{III} ions. As a result, the order of the thermal stability

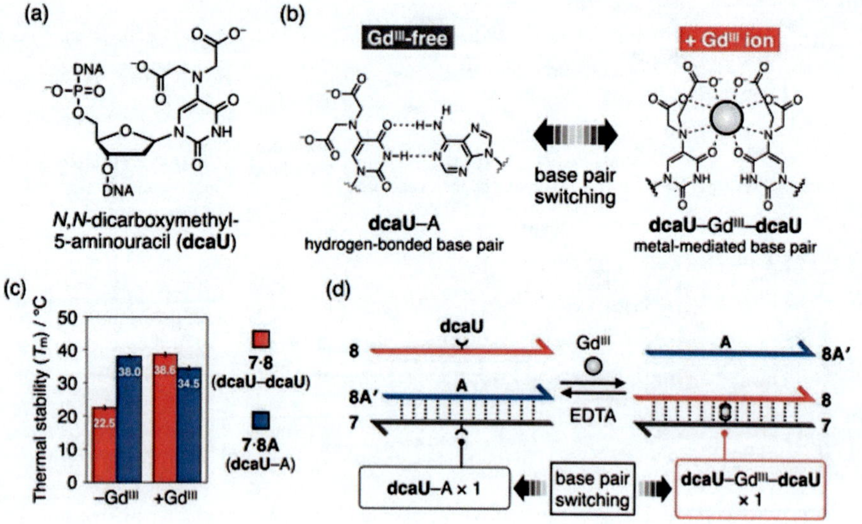

Fig. 5.3 **a** Structure of an *N*,*N*-dicarboxymethyl-5-aminouracil (**dcaU**) base. **b** Gd^{III}-dependent base pair switching of the **dcaU** base. **c** Melting temperatures (T_m) of DNA duplexes containing a **dcaU**–**dcaU** or **dcaU**–A pair in the absence and presence of Gd^{III} ions (1 equiv.). **d** Gd^{III}-dependent switching of the hybridization partner of **dcaU**-containing strand **7**

of the two duplexes was reversed in a manner similar to $\mathbf{U^{OH}}$, confirming the Gd^{III}-responsive base pair switching of the **dcaU** base (Fig. 5.3c). The base pair switching was applied to Gd^{III}-dependent control of the hybridization preference of **dcaU**-containing strand **7** (Fig. 5.3d). It was demonstrated that the hybridization partner of strand **7** was reversibly altered by the addition and removal of Gd^{III} ions.

Herein, coordination ability of the 5-modified uracil ($\mathbf{U^X}$) bases toward Gd^{III} ions was successfully improved by introducing the tridentate IDA ligand. As a result, the **dcaU** base behaved as a tetradentate ligand, forming octacoordinated metal complexes through 2:1 complexation with a Gd^{III} ion. This result supports that the coordination strength can be tuned by changing the 5-substituent of the $\mathbf{U^X}$ base. Accordingly, it is expected that the scope of metal ions of the metal-mediated base pairing would be expanded by carefully selecting the ligands for the modification.

In Chap. 4, I applied the base pair switching of $\mathbf{U^{OH}}$ to Gd^{III}-responsive control of DNA nanodevices. I selected DNA tweezers for the demonstration because they are one of the most studied DNA molecular machines exhibiting clear conformational switching. The $\mathbf{U^{OH}}$ bases were introduced into the arms of the DNA tweezers (Fig. 5.4). In the absence of Gd^{III} ions, self-assembly of four DNA strands yields the closed form of the tweezers. It was expected that the $\mathbf{U^{OH}}$-containing tweezers can be opened by the addition of Gd^{III} ions through destabilization of the $\mathbf{U^{OH}}$–A pairs and the intrastrand $\mathbf{U^{OH}}$–Gd^{III}–$\mathbf{U^{OH}}$ complexation. Based on native polyacrylamide gel electrophoresis (PAGE) and fluorescence measurement, I optimized the base sequences and solution conditions for efficient operation of the tweezers. Under

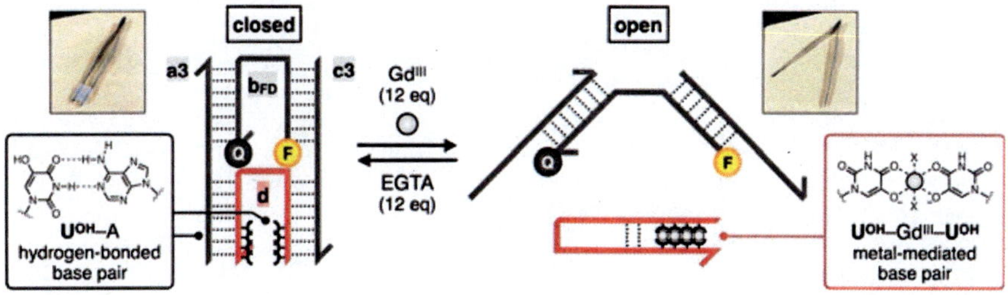

Fig. 5.4 Gd^{III}-responsive operation of DNA tweezers driven by base pair switching of the **U^OH** bases. "X" denotes additional ligands coordinating to the Gd^{III} ion such as H_2O and the neighboring nucleobases. F: FAM fluorophore, Q: Dabcyl quencher

the optimized conditions, the structural conversion of the DNA tweezers was demonstrated in response to Gd^{III} ions. Finally, I repeatedly switched the tweezer structures between the closed and open forms by the sequential addition of Gd^{III} ions and a chelating agent, EGTA.

In the operation of molecular machines or switchable nanodevices, it is important to control their structures and functions in a reversible manner. Since the metal complexation is inherently reversible, the base-pairing partner of the $\mathbf{U^X}$ base can be repeatedly switched by adding and removing the input metal ions. In fact, the $\mathbf{U^{OH}}$-containing DNA tweezers were iteratively switched between the closed and open states by the sequential addition of Gd^{III} ions and EGTA. Because of the efficient complexation and demetallation of $\mathbf{U^{OH}}$, the DNA tweezers were repeatedly operated without significant attenuation in the operation efficiency.

In summary, this study demonstrates the metal-responsive base pair switching of the 5-modified uracil ($\mathbf{U^X}$) bases and proves its applicability to dynamic DNA nanotechnology. The $\mathbf{U^X}$ base is broadly applicable to endow DNA assemblies with metal responsiveness because we can easily modify the 5 position of $\mathbf{U^X}$ with any metal-binding ligands. Therefore, it will be a useful molecular tool to expand the versatility of dynamic DNA nanotechnology in terms of the varieties of inputs and the programmability of the systems. Along with their great potential for the modification with various functionalities, the $\mathbf{U^X}$ bases are in good harmony with natural DNA structures since they retain the hydrogen-bonding site of the canonical nucleobases. Thus, the metal-responsive base pair switching of $\mathbf{U^X}$ is a highly unique tool which realizes both (i) structural and functional diversity and (ii) compatibility with natural biomolecular assemblies. In this context, this study is just the first demonstration of the applicability of the metal-responsive modified nucleobases. I believe that future studies will expand the versatility of the base pair switching, contributing to further development of dynamic DNA nanotechnology.

References

1. Seeman NC, Sleiman HF (2017) Nat Rev Mater 3:17068
2. Seeman NC (2016) Structural DNA nanotechnology. Cambridge University Press, Cambridge
3. Zhang F, Nangreave J, Liu Y, Yan H (2014) J Am Chem Soc 136:11198–11211
4. Zhang DY, Seelig G (2011) Nat Chem 3:103–113
5. DeLuca M, Shi Z, Castro CE, Arya G (2020) Nanoscale Horiz 5:182
6. Takezawa Y, Nishiyama K, Mashima T, Katahira M, Shionoya M (2015) Chem Eur J 21:14713–14716
7. Simmel FC, Yurke B, Singh HR (2019) Chem Rev 119:6326–6369

Curriculum Vitae

Keita Mori

JSPS Postdoctoral Fellow (CPD), Muraoka Group,
Department of Applied Chemistry, Graduate School of Engineering,
Tokyo University of Agriculture and Technology
2-24-16 Naka-cho, Koganei, Tokyo 184-8588, Japan
keitamori@m2.tuat.ac.jp
https://www.muraoka-lab.com/home-en
Orcid: 0000-0003-3572-7384

Education

- April 2014–March 2018
  B. Sc. in Chemistry, The University of Tokyo, Japan
  Supervisor: Prof. Mitsuhiko Shionoya

K. Mori, *Metal-Responsive Base Pair Switching of Ligand-type Uracil Nucleobases*,
Springer Theses, https://doi.org/10.1007/978-981-99-9400-7

- February 2017–March 2017
  Visiting Student in University of Cambridge, UK
  Supervisor: Prof. Jonathan R. Nitschke
- April 2018–March 2020
  M. Sc. in Chemistry, The University of Tokyo, Japan
  Supervisor: Prof. Mitsuhiko Shionoya
- April 2020–March 2023
  Ph. D. in Chemistry, The University of Tokyo, Japan
  Supervisor: Prof. Mitsuhiko Shionoya
- January 2022–April 2022
  Visiting Researcher in McGill University, Canada
  Supervisor: Prof. Hanadi F. Sleiman

Appointment

- April 2023–
  JSPS Postdoctoral Fellow, Tokyo University of Agriculture and Technology, Japan
  PI: Prof. Takahiro Muraoka

Research Interest

- Regulation of protein folding using synthetic molecules
- Dynamic control of biomolecular assemblies
- Redox chemistry inside biological cells
- Dynamics of metal ions in biosystems

Major Awards

- **Research Encouragement Award**, Graduate School of Science, The University of Tokyo, March 2023.
- **Next Generation Award**, 39th Medicinal Chemistry Symposium, November 2022.
- **ISNAC Outstanding Poster Award**, The 48th International Symposium on Nucleic Acids Chemistry (ISNAC 2021), November 2021.
- **CSJ Student Presentation Award**, The 101st Annual Meeting of the Chemical Society of Japan, March 2021.
- **Springer Nature Poster Award**, 15th International Symposium on Applied Bioinorganic Chemistry (ISABC15), June 2019.

Fellowships

- April 2021–March 2023
  JSPS Research Fellow for Young Scientists (DC2)
- January 2022–April 2022
  JSPS Overseas Challenge Program for Young Researchers
- April 2023–September 2023
  JSPS Research Fellow for Young Scientists (PD)